Natural Attenuation
CERCLA, RBCA's, and the Future of Environmental Remediation

Natural Attenuation
CERCLA, RBCA's, and the Future of Environmental Remediation

Patrick V. Brady, Ph.D.
Geochemistry Department
Sandia National Laboratories
Albuquerque, New Mexico

Michael V. Brady, J.D.
McDonough, Holland, and Allen
Attorneys at Law
Sacramento, California

David J. Borns, Ph.D.
Geophysics Department
Sandia National Laboratories
Albuquerque, New Mexico

LEWIS PUBLISHERS
Boca Raton   New York

Library of Congress Cataloging-in-Publication Data

Brady, Patrick V., 1961–
    Natural attenuation: CERCLA, RBCAs, and the future of environmental remediation
        p. cm.
    Includes bibliographical references (p.   ) and index.
    ISBN 1-56670-302-6 (alk. paper)
    1. Hazardous wastes--Natural attenuation.   I. Brady, Michael V.
II. Borns, David J.
TD1060.B73   1997
363.738'4--dc21
                                                                                            97-19742
                                                                                            CIP

This book contains information obtained from authentic and highly regarded sources. Reprinted material is quoted with permission, and sources are indicated. A wide variety of references are listed. Reasonable efforts have been made to publish reliable data and information, but the author and the publisher cannot assume responsibility for the validity of all materials or for the consequences of their use.

Neither this book nor any part may be reproduced or transmitted in any form or by any means, electronic or mechanical, including photocopying, microfilming, and recording, or by any information storage or retrieval system, without prior permission in writing from the publisher.

The consent of CRC Press LLC does not extend to copying for general distribution, for promotion, for creating new works, or for resale. Specific permission must be obtained in writing from CRC Press LLC for such copying.

Direct all inquiries to CRC Press LLC, 2000 Corporate Blvd., N.W., Boca Raton, Florida 33431.

© 1998 by CRC Press LLC
Lewis Publishers is an imprint of CRC Press LLC

No claim to original U.S. Government works
International Standard Book Number 1-56670-302-6
Library of Congress Card Number 97-19742
Printed in the United States of America 1 2 3 4 5 6 7 8 9 0
Printed on acid-free paper

Preface

This book analyzes the historical evolution and current direction of environmental remediation in the United States and outlines why there is now, and will be an increasing reliance on natural attenuation of hazardous substance toxicity in the coming years. Although the reasons for this shift in approach are legal, political, and scientific, each points in the same direction. The object of this book is to cover both the legal evolution and technical implications of remediation by natural attenuation. As a result the book moves at different speeds, depending on the perceived needs of the intended audience. The origins of CERCLA, sources of groundwater contamination, and groundwater hydrology are covered on a fairly basic level. There is a presumption that most environmental consultants know the finer points of groundwater hydrology, and that most environmental attorneys and regulators know how CERCLA works. The intent of the first three chapters is, therefore, to review the basics of CERCLA for the environmental consultants, and groundwater hydrology for the attorneys. The biogeochemical origins of natural attenuation are, instead, examined at a much more fundamental level in order to both explain and predict the macroscopic evolution of the cleanup effort. The nuts and bolts details of site assessment, remediation, and regulatory buy-in to the natural attenuation approach are then outlined at great length. Similarly, the shift in regulatory acceptance is analyzed in appreciable detail.

PVB wishes to acknowledge the support over the years of the National Science Foundation, DOE-BES/Geosciences, the U.S. Nuclear Regulatory Commission, the American Chemical Society, and Sandia National Laboratories. This work was supported by the United States Department of Energy under contract DE-AC04-94AL85000.

Patrick V. Brady
David J. Borns
Albuquerque, New Mexico

Michael V. Brady
Sacramento, California

The Authors

Patrick V. Brady is a geochemist at Sandia National Laboratories in Albuquerque, New Mexico. He received a Bachelors degree (Geology) at University of California — Berkeley, and M.S. and Ph.D. at Northwestern University (Geochemistry), and spent a year as a research environmental chemist at the Swiss Federal Institute of Technology. He then was an assistant Professor in the Geology Department and Adjunct Professor in the School of Engineering at Southern Methodist University. He has published approximately 30 scientific papers/book chapters/etc. on mineral weathering, global climate, acid mine drainage, water chemistry, low level radioactive waste disposal, and edited one book — *Physics and Chemistry of Mineral Surfaces* (CRC Press, Boca Raton, FL).

Michael V. Brady received his Bachelors degree in English at University of California — Berkeley, followed by a J.D. at Tulane University. He is an environmental attorney at McDonough, Holland, Allen in Sacramento, California, and specializes in the redevelopment of contaminated properties ("Brownfields"). He has represented a number of California counties, cities, redevelopment agencies, as well as private clients.

David J. Borns received his Bachelors degree from Dartmouth College, his M.Sc. from the University of Otago, NZ, and his Ph.D. from the University of Washington. His specialties are geophysical characterization of contaminated sites and the development of SMART geomembranes for landfills.

Acknowledgments

We thank Louise Fass (MH&A), who provided support and advice throughout; Harlan W. Stockman (Sandia), who helped write the section on irreversible sorption; and Jim Krumhansl (Sandia), who deutsch-uncled the whole thing from start to finish. Prof. Ron Dorn (ASU) gave us super pictures of lead in dead-end pores. Prof. Jeremy Fein (Notre Dame) provided the lead-in basalt example in Chapter 7. Pradeep Aggarwal, Neil Sturchio, and Chuck Douthitt gave us invaluable information about the perils of stable isotope measurements on biodegrading chlorinated solvents. Prof. Greg Bluth (Michigan Tech) gave us references on volcanoes and chlorinated hydrocarbons. K. L. Nagy provided helpful references on NORM. Charles Buckman and Susan Sorini gave us, and/or pointed us to, useful references on solids-leaching tests. Valda Terauds of Enhanced Solutions — Albuquerque, NM — graciously gave us her poster from the IBC Natural Attenuation Conference. For helpful reviews and/or support we thank Buddy Anderson (Sandia), W. D. Brady (LSU), Grace Bujewski (Sandia), Wu Ching Cheng (Sandia), L. T. Bryndzia, Leah Goldberg, Esq. (Hanson, Bridgett, Marcus, Vlahos, & Rudy — San Francisco, CA), Rod Hackler (Ecology and Environment — Chicago, Illinois), Frank Huang (NMTech), Diane Marozas (Sandia), Hans W. Papenguth (Sandia), Malcolm D. Siegel (Sandia), Marianne Walck (Sandia), and Hank Westrich (Sandia). Thanks also to Judy Neff (Sandia), Glenda Sweatt (Sandia), and Melissa Yoemans (VPI) for help with references. Our thanks are not to imply that the reviewers agree with all (or any) of the arguments presented here.

There was an old woman who swallowed a fly,
And I don't know why she swallowed the fly.
Perhaps she'll die.

There was an old woman who swallowed a spider,
That jiggled and wiggled and giggled inside her.
She swallowed the spider to catch the fly,
And I don't know why she swallowed the fly.
Perhaps she'll die … .

Contents

Chapter 1
Introduction 1

Chapter 2
The Law 7

Chapter 3
Sources of Hazardous Waste 15

Chapter 4
Groundwater Flow 31

Chapter 5
Chemical Attenuation 43

Chapter 6
Biodegradation 65

Chapter 7
Case Studies 81

Chapter 8
Demonstrating Natural Attenuation 121

Chapter 9
The Present 157

Chapter 10
The Future 167

Appendix 1
State Treatment of Natural Attenuation 173

Appendix 2
Glossary 179

Appendix 3
Excerpts from NPL Site RODs 183

Appendix 4
WWW Sources 217

References 221

Index 235

1 Introduction

Somewhere between 373 billion and 1.694 trillion dollars are projected to be spent in the next three decades to clean up hazardous waste sites in the United States of America (Russell et al., 1991). The bulk of this expenditure will be mandated by the Comprehensive Environmental Response, Compensation and Liability Act (CERCLA) and the Resource Conservation and Recovery Act (RCRA). Most of the money will be spent by the Department of Energy (DOE), the Department of Defense (DOD), other federal agencies, the various states, or at the direction of the Environmental Protection Agency (EPA). In 1993 the EPA estimated groundwater contamination was present at 2,000 CERCLA national priority list (NPL) sites, between 1500 and 3000 RCRA sites, 295,000 leaking underground fuel tank sites, 7300 DOD sites, 4000 DOE sites, 350 other federal facilities, and 20,000 state sites, for a total of roughly 330,000 sites all told (U.S. Environmental Protection Agency, 1993). The hundreds of billions of dollars slated for cleanup is money that cannot go to such areas as health care, education, social security, and infrastructure renewal. This might be defensible if the money were likely to provide an equivalent benefit to human or environmental health — but it won't. The actual health risk posed collectively by the sites is relatively minor (Milloy, 1995). Nor will the cleanup effort return most soils and groundwaters to anything approximating their precontaminated state. As pointed out by the National Research Council in 1994 "It is now recognized that for many sites, there are few if any technical remedies for meeting the goal of a permanent remedy" (National Research Council, 1994b). Freeze and Cherry (1989) spotted the general trend five years earlier and said "Attempts at aquifer remediation, where the goal has been to return a water supply to a state where its water meets drinking water standards, have almost without exception been a failure. If not an outright failure, they have made so little progress that expectation is for success not to be achieved in this century and maybe not in the next one." In fact, drinking water standards are commonly the nominal, but unachievable cleanup target. It is therefore impossible to demonstrate that the billions of dollars being spent every year will improve human or environmental health, now or in the distant future.

To find a way out of this seemingly intractable situation requires a wholesale re-examination of the various scientific and legal assumptions which led to the present state. Two legal thorns which cause expensive environmental cleanups are (1) the liability provisions of CERCLA (42 U.S.C.§§ 9601–9675), and (2) the use of MCLs (Maximum Contaminant Level Goals established under the Safe Drinking Water Act) as de facto cleanup targets. The first guarantees that at least half of the cleanup costs will be legal ones. The other ensures that a majority of the cleanups at the national, state, and local level will be treated like Superfund actions, which cost, on average, $27 million apiece (U.S. Environmental Protection Agency, 1993). The key scientific assumption that locks in high cleanup costs is that *contaminant availability in soils and groundwaters will remain constant unless actively remediated.*

In fact, toxicity is often cleansed from soils and groundwaters by natural processes ("natural attenuation") faster and more completely than by engineering approaches. EPA defines natural attenuation as "the biodegradation, dispersion, dilution, sorption, volatilization, and/or chemical and biochemical stabilization of contaminants to effectively reduce contaminant toxicity, mobility, or volume to levels that are protective of human health and the ecosystem". Reliance on natural attenuation is not a do-nothing ducking of responsibility; nor is it a walk-away solution. Remediation by natural attenuation generally requires considerable monitoring and results in contaminant reductions while providing tangible public health and environmental benefits. Unlike most present approaches, it doesn't just move contaminants from one place to another (with the attendant concerns and risks of transportation and ultimate disposal), but results in real reductions in contaminant mass.

The present system demands expensive, engineered solutions which, in general, don't work for a problem, which, judged by potential risks to human and environmental health, may not be much of a problem to begin with. A realistic consideration of the risk posed by the various sites would reduce considerably the scope of the remediation effort (Milloy, 1995). At the same time, explicit inclusion of natural attenuation in the cleanup decision-making process would cause a drastic reduction in cleanup cost, with or without CERCLA reform. The objectives of this book are therefore to:

1. Consider the legal and political forces which dictate cleanup actions,
2. Marshal the scientific evidence for natural attenuation of soil and groundwater toxicity.

NATURAL ATTENUATION

Over the past 3 decades or so there has arisen a belief that industrial by-products in the environment rattle around independent of natural processes until they are ingested, whereupon they shorten human lifespans and/or cause sterility. This assumption of potent and near-eternal toxicity, combined with the belief that extraction can be engineered, has caused the passing of unworkable and expensive environmental laws and regulations that have very little connection with the protection of human and environmental health, or for that matter, with geologic reality. The starting assumptions are for the most part false, though this has become clear only in the last few years. Many of the contaminants are degraded and/or isolated well before they have the potential to impact human health. Engineered extraction is rarely quick, and often unsuccessful.

Laws which depend on false diagnoses tend to grow in complexity as legal assumptions come into increasing conflict with reality. To compensate, lawmakers tend to adjust and complicate laws only enough to accommodate new complications instead of starting fresh with new models which conform to the true problem (Bobertz, 1996). CERCLA is no exception, and in 1995 the 104th Congress engaged in a wholesale revisitation of numerous of CERCLA's provisions. The 104th Congress just barely touched on one of the underlying premises of CERCLA — that a hazardous substance in the ground or groundwater is a hazardous substance that will cause trouble for the foreseeable future and must therefore be removed.

The false premise of all contaminants being waiting "time bombs" was popularized by Rachel Carson's book *Silent Spring*, wherein she showed that DDT, a pesticide used particularly after WWII, passed relatively unhindered from insects to their bird predators, which subsequently failed to reproduce adequately. The accumulation of DDT in raptors, the highest link in the insect food chain, was particularly acute in the United States before the banning of DDT in this country. Few quibble with the argument that the continued use of DDT led to severe repercussions for raptor populations in the post-war period. Indeed, raptor populations have rapidly recovered since DDT was phased out. Nevertheless, *Silent Spring* described certain truths about DDT which do not hold for the vast majority of the chemical compounds introduced into the environment over the past half century. Ames and Gold (1991) took issue with the statement from *Silent Spring* that "for the first time in the history of the world, every human being is now subjected to contact with dangerous chemicals, from the moment of conception until death". In fact we evolved in a wash of naturally occurring carcinogenic chemicals against a background of natural radiation. Plants manufacture pesticides to protect themselves from fungi, insects, and other predators, consequently the total quantity of natural carcinogens dwarfs the amount synthesized industrially (Ames and Gold, 1991). To quantify health risk it is critical that the health impact of new chemicals be gauged relative to the sizable background and the specific toxicity of the latter, not against an imagined, carcinogen-free Eden. If we want real solutions we must identify real problems.

The assumption that all contaminants are as focused as DDT in their ability to affect higher organisms is wrong. In fact, time starts to run out for most contaminants the moment they enter soils. Many hazardous chemicals have a skeleton of electron-rich organic carbon. Organisms that break down such compounds evolved very early in the history of the Earth, and have been breaking down organic chemicals far longer than we have been synthesizing them for industrial and agricultural use. Radioactive elements decay, and metals, to a greater or lesser extent, adsorb (stick) to soil minerals or precipitate to form their own insoluble crystals. When contaminants sorb, their potential for aqueous transport into and through the biosphere is severely limited. As transport is limited, the risk of exposure drops. For example, the radioactive isotope ^{137}Cs (a cesium atom with 137 protons and neutrons in its nucleus) released from Chernobyl, tends to stick to mineral surfaces, and appears to decay much faster than it becomes unstuck (desorbs). For this reason the health risk associated with it is minimized. Many metals and organics don't desorb at all over environmentally relevant time scales — the time required for them to decay, be biodegraded, and/or be diluted below regulatory limits. This arises from a number of causes. Contaminants that are bound somewhat loosely to mineral surfaces at first often diffuse into dead-end holes (microporosity) in mineral surfaces, and/or form progressively stronger bonds with the mineral surfaces. New mineral phases may also grow over and occlude sorbed contaminants. This surface sequestering may shield the contaminant molecules from degradation by microorganisms (in the case of organic molecules), but more importantly, it prevents their uptake into the biosphere.

Surface shielding therefore represents a net decrease in environmental toxicity, which is often measured to be 50% and more of the initial contaminant inventory.

Most contaminants in soils aren't waiting time bombs, but some are. Radioisotopes, and many organic contaminants, degrade at a rate proportional to their concentration, which means that their relative persistence can be treated in terms of their half-life, $\tau_{1/2}$, the amount of time it takes for a given quantity to decay to half its original amount. The half-life is specific to the contaminant, and, in the case of non-radioisotopes, depends on soil chemistry and biology. Many carcinogenic organic contaminants (e.g., chloroform, pentachlorophenol) are biodegraded, often to less toxic by-products, so rapidly in soils by indigenous organisms that they have half-lives of 6 months or less (Howard et al., 1991). It should be mentioned that the analogy between biodegradation half-lives and radioactive decay rates may not be completely valid at very low contaminant levels.

Some industrial chemicals don't just "go away" quickly (for example, DDT has a biodegradation half-life of roughly 20 years), and others, such as long-lived radionuclides, only "go away" over thousands of years. One of the objects of this book is to look specifically at the natural processes which dictate whether an industrial by-product is short-lived, and to be watched until it is degraded, or constitutes a long-lived hazard which warrants active environmental remediation. Another object of this book is to focus the debate on those situations where neither natural attenuation nor environmental remediation works. The monitor-or-remediate decision should always be based on a *persistence-to-distance* judgment. Specifically, the time required for a toxin to cover the distance between the source and potential receptors should always be compared to the rate at which the toxin is degraded along the way. If natural processes reduce contaminant levels to safe levels over short distances, active remediation is uncalled for. If, because of slow degradation, fast transport, and/or close receptors, contaminant levels are likely to remain unchanged, remediation efforts, if they can be proven to be effective, should be considered.

The stated objective of CERCLA is to minimize risk to human and environmental health. Implementation of Superfund cleanups take, on average, 2 years. Yet, to reach the implementation stage at present, requires, on average, 10 years. For a carcinogen with a soil half-life of a year, the 10-year wait from identification to implementation constitutes at least 10 half-lives. 10 half-lives means a reduction of the contaminant to roughly one thousandth of its original level. In essence, doing nothing but the Superfund paperwork, i.e., litigating and waiting for 10 years can give a better than three order of magnitude drop in toxicity (e.g., Borgert et al., 1995). This may be enough to get contaminant levels down to drinking water levels. At the same time, a factor of 1000 reduction almost always dwarfs the amount of attenuation achievable by engineered methods such as pump-and-treat (see below).

A number of natural soil processes cause similar reductions in contaminant levels. But some times they do not. A reasonable way to determine which will, and which won't, is to examine the mechanistic origins of contaminant attenuation in soils. These are chemical, physical, and biological, often coupled, and therefore somewhat complex. Although the fine-scale features of each of these areas remain the foci of ongoing research, many of the most important order-of-magnitude controls on contaminant attenuation are already fairly well understood, and somewhat predictable (see e.g., Dragun, 1988). Their effect on contaminant transport can therefore be effectively assessed. An incomplete understanding of the fine-scale details of the various attenuation pathways should not prevent reliance on this method of site remediation.

Quantification of natural attenuation might be even more critical for assessment of ecological risk, which is otherwise decided in the absence of numerical targets and guidelines (Milloy, 1995). Natural attenuation of contaminants is arguably a rough litmus test of the resilience and self-cleansing capacity of exceedingly complex soil systems. Numerical assessment of natural attenuation might therefore provide an approximate measure of the specific ecological risk posed by a contaminant.

Presently, soil/groundwater cleanups are performed in order to reduce the levels of chemicals in the ground, and, incidentally, to reduce risks to people. In effect, persistence is assumed infinite, and expensive, engineered solutions are generally brought to bear. A more effective, and certainly cheaper, approach is to try to remove only those contaminants that natural soil processes won't. Unfortunately, in many situations existing remediation technologies are unable to assure that the desired levels of cleanup can even be reached. Remediation is rarely cheap and the often poor prospects for final success make it critical that the cleanup target be chosen wisely. The question of "how clean is clean" should be decided based on health risk and environmental transport factors and not on a desire for zero contaminant levels.

We didn't evolve on a toxin-free Earth. Regulatory targets or cleanup efforts which shoot for this imagined situation are naive and incredibly expensive. Cleanup activities should aim for endpoints which, once achieved, represent a tangible contribution to human and environmental health. Cleanup efforts, because they are expensive and because they often produce their own health risks, shouldn't be done if a real health benefit cannot be demonstrated. Environmentally acceptable endpoints can only be defined on a scientific basis.

Chapter 2 of this book examines the mechanics of hazardous substance regulation and how it dictates cleanup strategies and costs. Chapter 3 outlines the types of hazardous wastes commonly found in soils and groundwaters. Chapters 4, 5, and 6 explore the fundamental origins of natural attenuation and outline the physical, chemical, and microbiological controls on chemical reactivity. There is an emphasis in each of these chapters on modeling distance (Chapter 4) and persistence (Chapters 5 and 6). In Chapter 7 a number of case studies are used to emphasize the self-cleansing capacity of soils, and to point out just what should, and should not, be cleaned up from a natural attenuation perspective. Chapter 8 presents a number of protocols which have been advanced for demonstrating natural attenuation. Note that we deal here with contamination existing in soils after the source has been removed. Superfund has two components; removal of immediate public health threats, and remediation of what's left behind. The first generally costs around $2 million per site, takes less than a year and usually eliminates human exposures to site substances (Milloy, 1995). It is the later remediation step on which we focus.

2 The Law

The universe of hazardous substance cleanups can be divided into three categories: NPL sites, federal facility sites, and all others. The first category, which includes all EPA supervised cleanups of sites on the National Priorities List, is actually quite small (roughly 1300). Though EPA promises to expand this list, current congressional opposition suggests that the list is not going to be expanded any time soon. In fact, the 104th Congress attempted to prohibit further additions to the NPL.

The second category consists of federal facility sites administered by the United States Departments of Energy and Defense. EPA's involvement at these sites is limited, though all cleanup is to be performed by the responsible federal agency subject to cooperative agreements with EPA. There are more of these sites than NPL sites, and the problems for the DOE are often more complicated technically because they often contain mixed radioactive and non-radioactive hazardous waste.

The final category includes all other cleanups performed by private parties under state, EPA or judicial supervision. The problems at these sites are more often more tractable than the problems at NPL and federal sites; however, the number of these sites dwarfs that of the other two categories.

What happens at each site, no matter what category it occupies, is governed ultimately by the provisions of CERCLA and the federal agency charged with implementing CERCLA — EPA. Although EPA has direct day-to-day responsibility for only a small number of cleanups. The majority of cleanups are impacted by regulations promulgated by EPA for use with NPL sites. To understand this distinction one must understand CERCLA.

CERCLA HISTORY

CERCLA was passed in December, 1980, in the waning days of the Carter administration and originally consisted of a number of hastily drafted "solutions" to the problem of hazardous waste sites. Coming two years after the Love Canal incident there was little debate and less analysis. CERCLA was subsequently amended by the passage of the Superfund Amendments and Reauthorization Act (SARA) in 1986. (We use the term CERCLA to include SARA unless stated otherwise). Among the current provisions of CERCLA are

1. A liability scheme applicable to all sites.
2. Money to pay for cleanups at the "worst" sites where a responsible party cannot be identified, i.e., those sites on the NPL.
3. The National Contingency Plan which sets out management principles and a remedy selection process.

CERCLA's liability scheme is incredibly broad. It imposes liability, strictly, jointly and severally upon current owners of the contaminated property, owners of the property at the time of disposal, transporters of hazardous substances to the contaminated property and arrangers for disposal of hazardous substances at the property. The liability is also retroactive. The courts have interpreted these already broad provisions even more broadly.

CERCLA's liability provisions in the abstract are frightening enough for potentially liable parties, but they grow more alarming when one considers the average 27 million dollar price tag of the cleanups. It is little wonder that upwards of 50% of the transactional costs of a typical CERCLA cleanup go to attorneys' fees. This figure is brought up in nearly every criticism of CERCLA; however, the legal opposition purchased by these fees is logically inescapable given CERCLA's liability scheme. For all but the most affluent business concern, voluntary acceptance of cleanup responsibility means bankruptcy. As a matter of straight economics, most would rather fight in court than commit financial suicide. On average, the ~ 50% that doesn't go to attorneys does not go to cleanup either; 38% goes to support the federal, state, and local regulatory apparatus, and only 12% actually pays for cleanup.

CERCLA's liability scheme is the primary target of those who would amend the statute, yet this book focuses on the other 50% — the non-legal costs. Cleanup costs can be dramatically lowered by changing the method of remedy selection, without CERCLA liability reform, and legal costs will rapidly follow.

As long as remediation costs remain as high as they currently are, no politically palatable reform of the liability scheme is likely to persuade those who share some responsibility for contamination to cooperate instead of litigate. Put simply, if the stakes are high, the attorneys' fees will be high.

CERCLA MECHANICS

Who determines how the other 50% (the non-attorney portion) is spent? To a certain extent, EPA does, but it is not that simple. EPA directly supervises cleanups at NPL sites but not at federal sites or at the numerous non-NPL sites nationwide. Federal sites are generally remediated by the responsible federal agency consistent with the National Contingency Plan (NCP) — see below. At non-NPL, non-federal sites cleanups are generally performed consistent with the NCP, without EPA supervision yet in the shadow of CERCLA liability. As will be explained later, this policy results in an expensive "worst of both worlds" situation for the majority of sites in the United States. But before looking at how the NCP is an expensive and wasteful proposition in the absence of EPA supervision, one must first understand the source of the original error, and for that one must understand how EPA determines cleanup standards at NPL sites.

One of the original provisions of CERCLA called for the revision and republication by EPA of "The National Contingency Plan for the Removal of Oil and Hazardous Substances" (42 U.S.C. § 9605). This was done because the NCP was originally prepared and published by EPA in 1973 under the authority of Section 1321(c) of the Clean Water Act. The NCP was amended and incorporated into CERCLA in 1980 and has been amended three times since then, in 1982, 1985, and 1990. As originally written, the NCP was to contain a national hazardous substance response plan for dealing with releases of hazardous substances, which at a minimum included:

1. Methods for discovering and investigating facilities where hazardous substances had been disposed;
2. Methods for evaluating and remedying releases;
3. Methods and criteria for determining the appropriate extent of cleanup (42 U.S.C. § 9605(a)(1)-(3)).

Based upon the criteria identified in the NCP, EPA was to identify and list the one hundred highest priority sites in the country (42 U.S.C. § 9605(a)(8)(B)). This list was the original NPL. As a side note, Section 105 of CERCLA, when describing what EPA should include in the NCP, reflected the political nature of the process as the NPL was to include at least one site from each of the fifty states (42 U.S.C. § 9605 (a)(8)(B)). The same section testifies to the scientific naivete at the time in stating that all cleanups would be completed within 10 years of initiation (42 U.S.C. § 9604(e)(s)).

Cleanup of sites on the NPL was required to be, "to the extent practicable," consistent with the NCP (42 U.S.C. § 9621), which in itself is not too disturbing because the NCP does not contain cleanup standards. Section 121(d) of CERCLA does; kind of. It states: "Remedial actions selected under this section ... shall attain a degree of cleanup of hazardous substances, pollutants, and contaminants released into the environment and of control of further release at a minimum which assures protection of human health and the environment. Such remedial actions shall be relevant and appropriate under the circumstances presented by the release or threatened release of such substance, pollutant, or contaminant" (42. U.S.C. § 9621(d)(1)).

This quote does not provide a standard either, but it is closer. A subsequent provision of this section finally arrives at a numerical standard stating that with respect to remedial actions that result in hazardous substances left on site, "Such remedial action shall require a level or standard of control which at least attains Maximum Contaminant Level Goals (MCLs) established under the Safe Drinking Water Act — 42 U.S.C. Section 300f et. seq."

In other words, cleanups at NPL sites must achieve MCLs. Indeed, drinking water standards were the cleanup goal for 90% of the cleanups done between October, 1987 and September, 1991 (MacDonald and Kavanaugh, 1994). Therein lies the problem. The Safe Drinking Water Act was passed to ensure that the water coming out of your faucet was safe to drink, not to provide a cleanup standard for hazardous waste sites, or for that matter the standard for the water in the local swimming pool. MCLs are set through the performance of exposure assessments which are intended to arrive at a numerical limit beyond which exposure may impair health. By their very nature, exposure assessments exaggerate the risk of a given contaminant. For exposure assessments, Milloy (1995) cites the following assumptions that often tend to exaggerate the calculated risk:

1. Site groundwater is potable (At NPL sites it seldom is).
2. The highest measured, or calculated upper-bound, contaminant levels are present site-wide (rarely).
3. Groundwater values are total, not dissolved, concentrations from unfiltered samples.
4. Contaminant levels remain constant throughout the exposure period.
5. Groundwater is ingested at a rate of 2 liters per day (Not from NPL sites it isn't).
6. Individuals will consume 100% of their drinking water from contaminated wells, and will do it for 30 years (Again, unlikely).
7. Individuals will shower with 50 gallons of water for 20 minutes per day in a 3' by 5' shower stall and then will spend another 10 minutes in a 5' by 9' bathroom.

Most of the assumptions covering dose mortality curves, shower stall dimensions and the like are outside the scope of this book (instead — see Milloy (1995)); assumption 4 is not. In the chapters that follow we will outline the reasons why contaminant levels almost invariably decrease with time, and how neglect of this leads quite often to a very large exaggeration of risk. But before we do, we must emphasize why the reliance on MCL targets, confined by its terms to cleanups at the worst sites in the country, is such a pernicious influence at all the other cleanups.

The reason is two-fold. First, from the liability standpoint, CERCLA's liability scheme applies to NPL and non-NPL sites equally. In other words, the risk of CERCLA liability exists wherever there is a release of a hazardous substance into the environment, with the exception of petroleum products which are exempted from CERCLA's reach. Second, most of the state Superfund laws passed in the 1980s and 90s incorporate by reference the requirement that the cleanup be "consistent with the NCP" which in turn incorporates the cleanup standards from the Safe Drinking Water Act, i.e., MCLs.

The non-NPL sites see the worst of both worlds in the flexibility that EPA has in choosing cleanup levels. The NCP provides for use of cleanup goals other than MCLs, but, as a practical matter, to use alternative cleanup goals a great deal of money (or none at all) is needed to persuade EPA that it is wise to do so in a given situation (See Appendix 3 which summarizes a number of the NPL sites where natural attenuation has been selected as the final remediation method). Politically speaking, EPA occupies the high ground when determining cleanup goals, and choosing cleanup goals different than the MCLs is politically risky because it can too easily be painted as "making deals with polluters." The MCLs thus become the norm.

On the non-NPL level where there is no EPA supervision, there is no flexibility. State laws require consistency with the NCP, but the state environmental agencies seldom have the ability to apply the flexible cleanup goals allowed by the NCP even if they were willing to do so. It gets worse. In the purely private setting, where the owner of contaminated property is suing other responsible parties for his cleanup costs under CERCLA, the owner has the affirmative obligation to show that his/her cleanup was "consistent with the NCP." If the owner chooses alternate cleanup goals the owner risks a complete recovery. EPA points out accurately, but rather unhelpfully, in the federal register explaining the 1990 NCP, that parties do not *have to* conduct cleanups consistent with the NCP. It is only if the responsible party wishes to file a CERCLA action that the remediation must be consistent with the NCP (55 FR 8796). Without the benefit of CERCLA's broad cost recovery provisions, few responsible parties would be willing to even consider cleaning up a property voluntarily. CERCLA's liability scheme, when paired with its remedy selection method, compels unwise economic choices, i.e., the most conservative cleanup method possible, no matter how inefficient or expensive.

NATURAL ATTENUATION AND CERCLA

In CERCLA and the NCP natural attenuation is a bit player. In the real world natural attenuation often plays the lead role. This discrepancy is a result of the well-intentioned political forces which drove the enactment of CERCLA and an incomplete scientific understanding of natural attenuation in 1980 and 1986 when CERCLA was before Congress.

There is no mention in CERCLA of natural attenuation, but there is in the current NCP which expands upon the statutory criteria and provides more guidance on the remedy selection method. The NCP, and the federal register that explains the NCP's terms, essentially sets forth the scope of EPA discretion and suggest how EPA will exercise that discretion in selecting and implementing a remedy. The NCP is clear in its bias towards active remediation, i.e., engineering approaches like pump-and-treat systems for contaminated groundwater. Such systems are not only quite expensive but, as explained later, they are also seldom successful (National Research Council, 1994b). As pointed out in 1994 by the National Research Council, "there are limits to what technology can accomplish and existing regulatory requirements do not adequately account for these limits" (National Research Council, 1994a).

When the 1990 NCP was in draft form and circulated for public comment, a number of commenters argued for and against the use of natural attenuation for remediation of groundwater (55 Fed.Reg. 8733-8735). On one side, some commented that EPA should rely more on natural attenuation in its cleanup method selection process, particularly where there had been earlier source removal. On the other side, commentors correctly pointed out that some contaminants do not readily degrade in the subsurface. EPA gave a tepid endorsement of use of natural attenuation in a few very limited circumstances. Because the selection of a remedy under CERCLA is within the exclusive jurisdiction of EPA, the tone of the 1990 NCP suggests that natural attenuation would rarely be allowed by EPA. Recent history is somewhat encouraging, with natural attenuation providing the primary remediation method for closure of an increasing number of NPL sites in the last four years. When looking through Appendix III, which explains those sites where natural attenuation has been approved, an either/or approach to natural attenuation begins to appear. In a few cases, there is a recognition of the effectiveness of natural attenuation in erasing soil toxicity. In other cases, though, natural attenuation is grudgingly accepted only when the responsible parties refuse to fund more expensive engineering exercises (e.g., B & B Chemical Co. Inc.), or when there are no solvent responsible parties available (e.g., Holton Circle). If there is any rhyme or reason to EPA's selection of natural attenuation as a remedy in a given situation, the cases in Appendix III suggest that the selection has more to do with who is the responsible party and how much money is available. There are very few contaminants that EPA has determined can *not* be naturally attenuated.

A fundamental shift in regulatory acceptance may have been started with a signal from Lawrence Livermore National Laboratory which issued its "Recommendations to Improve the Cleanup Process for California's Leaking Underground Fuel Tanks," on October 16, 1995. The basic conclusion of the Lawrence Livermore Report was that migration of petroleum hydrocarbons in soil and groundwater beneath a majority of leaking underground fuel tank sites in California was severely limited by natural attenuation (see below). The unstated conclusion was that the California Environmental Protection Agency (CalEPA) was wasting a lot of money over-regulating and over-remediating an otherwise relatively minor problem. The regulatory response was grudging acknowledgment of the science but rejection of the use of passive bioremediation in favor of "state of the art methods and technology," such as soil vapor extraction/bioventing and air sparging/bio-sparging. At the time of writing the California Legislature was taking matters into its own hands and forcibly injecting the Lawrence Livermore study into the underground storage tank laws. Note that the Lawrence Livermore study was not the first point at which the scientific community recognized natural attenuation of petroleum hydrocarbons. The process was fairly well understood and worked out in the late 1980s and early 1990s (see below).

Ritz (1996) examined the acceptance of natural attenuation at the state level. The most accurate statement that can be made about state policies regarding natural attenuation is that almost all are in the process of being changed. At the writing of Ritz's report (late-1996) 38 states were reviewing their positions on natural attenuation with an eye on changing them (see Appendix 1).

It is easy to blame the move towards natural attenuation on shifting political sands and label the net results as "environmental retreat". In fact, politics doesn't have a whole lot to do with it. What is driving the shake-up in environmental cleanup is simple. The staggering costs have forced a re-examination of how the cleanups are performed. At the same time there has been a slow accumulation of case histories demonstrating that groundwater is often better cleaned by natural processes. Further examination, in many cases, also reveals that the health risks just don't merit the cash outlay of engineered remediation.

In Washington, as in Sacramento, the legislative branch appears to be setting the agenda in the face of regulatory opposition to natural attenuation. The 104th Congress' attempts to reauthorize CERCLA in the 1995 legislative session would have removed the bias towards quick engineering fixes and would have included explicit statutory recognition of natural attenuation. The proposed legislation would also have ended EPA's practice of treating all waters as potential drinking water sources. The 105th Congress has taken up where the 104th left off, targeting the same shortcomings for overhaul.

CONCLUSION

EPA's methodology for choosing a cleanup strategy makes sense only if viewed as the regulatory reflection of an overly optimistic statutory cleanup regime. This methodology is biased towards active, expensive remediation technologies that rarely work. The methodologies are tied to MCLs because CERCLA requires it, and because Congress didn't know any better in 1980. The law was drafted assuming sites *had* to be cleaned up in order to protect public health and that sites *could* be cleaned up. Improvements in cleanup technology were assumed, and then relied on, to take up the slack between legislative desire and reality. For the most part it hasn't happened. No "magic bullets" have been invented to drastically boost cleanup efficiencies in the 16 years since CERCLA was passed. Pumping and treating remains the tool of choice at three quarters of the Superfund sites with contaminated groundwater, as well as at most sites governed by RCRA and state laws (MacDonald and Kavanaugh, 1994). It is foolish to assume that the cleanup tools used in the next thirty years are going to be drastically more efficient than the ones we've employed for the past 16 years (see, e.g., Freeze and Cherry, 1989).

We clearly cannot buy clean aquifers, though that hasn't stopped us from spending an enormous amount of money to prove the point. The seeds of CERCLA's failure were (1) a poor understanding of natural processes, and (2) an over-optimistic expectation of technological advance. We propose that a more sober examination of the natural soil processes alone, will, in fact, point to a methodology for achieving CERCLA's goals, that is more defensible from a scientific and public health standpoint, and that is substantially cheaper as well.

BACKGROUND READING

MacDonald J. A. and Kavanaugh M. C. (1994) Restoring contaminated groundwater: An achievable goal? *Environ. Sci. Technol.* (28) 362-368A.

Milloy S. J. (1995) *Science-based Risk Assessment.* National Environmental Policy Institute. Washington, D.C.

3 Sources of Hazardous Waste

INTRODUCTION

A chemically complex and voluminous waste stream goes with the territory of being an industrialized country, and the sources of wastes are many and diverse. The U.S. mining industry generates upwards of 2 billion tons of solid waste a year, fully 40% of the total volume of waste (Hoye and Hubbard, 1989). This consists primarily of overburden (the rock on top of the ore body), tailings (ore processing wastes), and leach residues, generated by the chemical extraction of ore elements from the rock. Household waste is generated at a rate of about 20 lbs per person per week, which, nationwide, amounts to ~125 million tons per year. 291 million tons of waste landed in municipal solid waste landfills in 1988, though much of this consisted of construction and demolition debris, yard waste, and the landfill material used to bury the waste periodically (Alexander, 1993). Storage tanks beneath gasoline stations leak. Lead, used for decades as a gasoline additive, now resides in soils world-wide where it might encounter ^{241}Pu (Pu = plutonium), and its decay product ^{241}Am (Am = americium), which rained down from above-ground testing of nuclear weapons in the 40s, 50s, and 60s. Most people in the U.S. can also find ^{241}Am closer to home, in their automatic smoke detectors. Hospitals generate waste to the tune of approximately 20 lbs per day per bed, 10% of which is infectious, and ~21.4 mL of hazardous waste per bed per day (Cross and Robinson, 1989). The latter seems small, but amounts to more than a ton a year for a 200-bed hospital (Cross and Robinson, 1989). Hazardous wastes are generated in large quantities by households using over-the-counter cleaners, polishes, solvents, paint thinners, paint strippers, adhesives, herbicides, and pesticides, to name a few. The integrated quantity is not trivial. For example, a 1983 survey in Albuquerque, New Mexico showed that every 100 households generated a little less than a ton of hazardous waste per year. Enormous volumes of hazardous waste are produced during the refining of petroleum and the synthesis of industrial chemicals. Table 3.1 outlines the more common chemicals which appear at hazardous waste sites.

The chemical characteristics which determine the fate and transport of the various substances in Table 3.1 in the environment differ widely, and deserve close examination. This is best done by dealing with the inorganic and organic materials separately.

METALS

Before proceeding it must be noted that toxicity depends not merely on the presence of a particular substance, but rather on the actual amount. All of the metals listed

TABLE 3.1
The Most Frequently Detected Groundwater Contaminants at Hazardous Waste Sites

Rank	Compound	Common sources
1	Trichloroethylene	Dry cleaning; metal degreasing
2	Lead	Gasoline; mining; pipes
3	Tetrachloroethylene	Dry cleaning; metal degreasing
4	Benzene	Gasoline; manufacturing
5	Toluene	Gasoline; manufacturing
6	Chromium	Metal plating, cleaning agent
7	Methylene chloride	Degreasing; solvent
8	Zinc	Manufacturing; mining
9	1,1,1-Trichloroethane	Metal and plastic cleaning
10	Arsenic	Mining; manufacturing, cemetaries
11	Chloroform	Solvents
12	1,1-Dichloroethane	Degreasing; solvents
13	1,2-Dichloroethene	Breakdown product of trichloroethene
14	Cadmium	Mining; plating
15	Manganese	Manufacturing; mining
16	Copper	Manufacturing; mining
17	1-1-Dichloroethene	Manufacturing
18	Vinyl chloride	Plastic and record manufacturing, and breakdown product of chlorinated solvents
19	Barium	Manufacturing; energy production
20	1,2-Dichloroethane	Metal degreasing; paint removal
21	Ethylbenzene	Styrene and asphalt manufacturing; gasoline
22	Nickel	Manufacturing; mining
23	Di(2-ethylhexyl)phthalate	Plastics manufacturing
24	Xylenes	Solvents; gasoline
25	Phenol	Wood treating; medicines

Note: The ranking reflects the number of sites at which the substance was detected in groundwater.

From National Research Council, 1994b.

in Table 3.1 exist naturally in detectable amounts in soils, and some for that matter, show up in vitamin and mineral supplements. Lead, chromium, arsenic, barium, beryllium, silver, cadmium, zinc and manganese are the metals of primary environmental concern. Figure 3.1 gives an idea of the metal wastes of concern to DOE. Note though, that although it appears in Table 3.1, manganese is an essential nutrient for plants and animals, as are zinc, copper and chromium. While high levels of Zn and Cu aren't particularly bad for humans at low levels, both are detrimental to aquatic life when present at very low concentrations. Cu and Zn are present in at least minor quantities in most ores mined for metals. Consequently, when water leaches through the mines or associated tailings piles, there is a chance that toxic levels will make it into streams and rivers where they can seriously affect aquatic populations.

Sources of Hazardous Waste

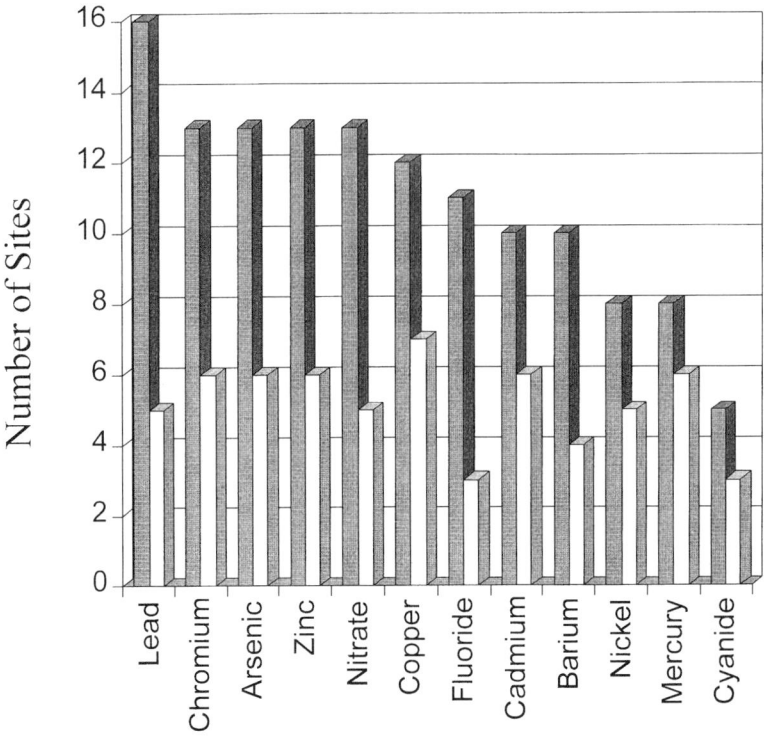

FIGURE 3.1 Number of DOE sites contaminated with metals. Gray bar, groundwater; open bar, soils. (From Riley et al., 1992).

CADMIUM

Cadmium (Cd) is used in plating operations, to make nickel-cadmium batteries and PVC pipe, and as a coloring agent. Cadmium is a trace component of phosphate soil amendments (e.g., fertilizers), and lead and zinc ores. Application of one and/or smelting of the others can therefore lead to the dispersal of cadmium into the environment. Cadmium is chemically similar to zinc. Substitution for the latter in metabolic processes leads to a variety of deleterious health effects, such as high blood pressure, kidney damage and red blood cell loss. In nature, cadmium exists as the divalent cation, Cd^{2+}. Below pH 8 Cd^{2+} generally exists as the free ion. At higher pH, Cd^{2+} drops from solution to form $Cd(OH)_2$ and/or $CdCO_3$ (otavite), or is sorbed onto various soil solids. Higher levels of Cd can be maintained in solution given adequate levels of sulfate (SO_4^{2-}) and/or chloride (Cl^-), both of which form ion pairs with Cd^{2+}. Ion pairing is the association of two dissolved chemicals in solution. Under the oxygen-poor conditions found in bogs and water-logged soils and sediments Cd forms relatively insoluble cadmium sulfide. It should be pointed out here that solubility is a relative term. If a chemical is insoluble, very little of the chemical remains in the dissolved state. Water in contact with a soluble mineral, such as table salt, carries a very large amount of the mineral component dissolved

TABLE 3.2
Federal Drinking Water Standards

Contaminant	MCL (mg/l unless noted)
Arsenic	0.05
Antimony	0.006
Barium	2
Beryllium	0.004
Cadmium	0.005
Chromium	0.1
Cyanide	0.2
Fluoride	4
Mercury	0.002
Nickel	0.1
Nitrate (as N)	1
Selenium	0.05
Organics	
Benzene	0.005
Carbon Tetrachloride	0.005
1,1,1-Trichloroethane	0.2
TCE	0.005
Vinyl Chloride	0.002
Xylenes (total)	10
PCBs	0.00005
PCE	0.005
Toluene	1
Ethylbenzene	0.7
1,1,2-Trichloroethane	0.005
trans-1,2-Dichloroethylene	0.1
cis-1,2-Dichloroethylene	0.07
1,1-Dichloroethylene	0.007
1,2-Dichloroethane	0.005

Note: Standards as of 1994.

in solution. Cadmium can also combine with arsenate, chromate, or phosphate to form insoluble compounds. The MCL for Cd is 5 parts per billion (see Table 3.2); 1 part per billion = 1ppb = 1μg/L. 1ppm = 1000 ppb.

LEAD

Lead (Pb) is used in storage batteries, but its environmental impact comes largely from utilization in pigments and as a gasoline additive (tetraethyllead — see Figure 3.2) to improve combustion (it also reduces valve wear). The latter source, as well as the production of lead during smelting and the burning of coal, has led

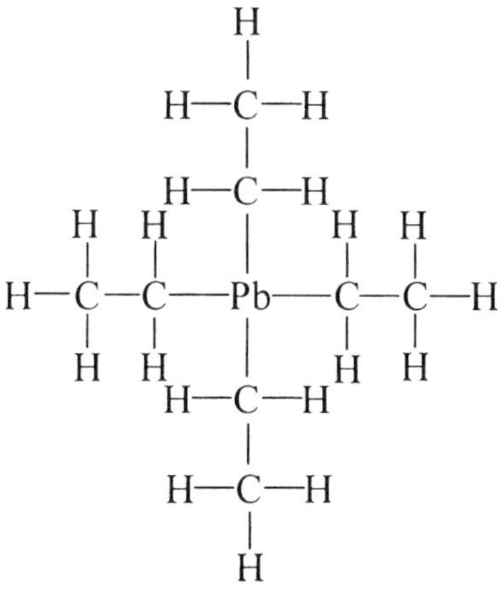

Sources of Hazardous Waste

```
            H
            |
          H—C—H
            |
          H—C—H
   H  H    |    H  H
   |  |    |    |  |
 H—C—C—Pb—C—C—H
   |  |    |    |  |
   H  H    |    H  H
          H—C—H
            |
          H—C—H
            |
            H
```

FIGURE 3.2 Tetraethyllead.

to a near-global distribution of lead. Lead exists in solution as Pb^{2+} which forms aqueous complexes with chloride, and hydroxyls at high pH. Given the appropriate conditions, Pb^{2+} also forms ion pairs (aqueous complexes) with sulfate and carbonate. Lead is, by and large, very insoluble and forms hydroxycarbonate minerals in most soils where it is present (Hem 1989). Moreover, lead sorbs to many soil mineral surfaces (Hem, 1976b). Solution lead levels are almost always low, reflecting the tendency of lead to stay in, or on, solids. The primary toxicity pathway is then ingestion of the latter. Pb-containing soils and paint chips are obvious items which young children, in particular, tend to ingest. Lead can also be picked up if drinking water moves through leaded pipe. Lead in the body inhibits the formation of hemoglobin, damages kidneys, and leads to IQ deficits and mental retardation. The mean residence time of lead in blood is 35 days, and the majority (75% for children and 95% for adults) resides in the bones. Young children absorb 40 to 50% of ingested lead. For adults, this number is 10 to 15% (Renner, 1995). Lead, once ingested, is eliminated from the body very slowly. Lead levels have dropped in children in the U.S., as has our estimate of the level thought to cause deleterious health effects.

MERCURY

Mercury (Hg) was used as an early fungicide, and in electrolytic processes. Mercury in the brain causes depression, irritability, and madness. Prolonged exposure to mercury (as well as arsenic and/or lead) may lead to peripheral neuropathy; numbness and a "pins and needles" sensation. Mercury was used to tan hides, and the uptake of Hg by hat makers in the 19th century gave rise to the term "Mad Hatters". Mercury is a common industrial waste which is widely dispersed in the environment due to its volatility. Several thousand cubic meters of DOE's radioactive waste is contaminated with mercury as well, complicating the ultimate extraction and/or treatment of the radioactive component. Burning of coal provides a significant input of mercury into soils and waters (some is also emitted by volcanoes and geothermal activity). Mercury can exist in soil and groundwaters in a variety of valence states. Elemental mercury (quicksilver) is stable, though in general mercury exists as Hg^{2+}. Over wide pH ranges, Hg^{2+} attracts hydroxyls, and/or chloride ions. These lower-charge ion pairs sorb negligibly to most minerals. Methyl mercury, $HgCH_3^+$, formed in the presence of methane-producing bacteria in natural waters, has a tendency to concentrate in the tissue of aquatic animals. Uptake of ingested methyl mercury is on the order of 100%; for non-methyl mercury the uptake is on the order of 10%. Mercury causes kidney damage as well. The MCL for mercury is 2 ppb.

CHROMIUM

Chromium (Cr) is used in metal plating and as a cleaning agent. Chromium exists in two charged (valence) states in water; Cr^{3+}, which is quite insoluble under most conditions and only weakly toxic, and Cr^{6+}, which is soluble and highly toxic. Prolonged exposure to chromate leads to bronchogenic cancer, as well as the formation of cranial abscesses. Hexavalent Cr exists in solutions of pH >6 primarily as CrO_4^{2-}. Below pH 6 it exists as $HCrO_4^-$, and generally to a lesser extent, as $Cr_2O_7^{2-}$. At very high chromate levels the latter species becomes more abundant. Trivalent Cr exists as Cr^{3+}, $Cr(OH)^{2+}$, and $Cr(OH)_2^+$ in most soil solutions. The insoluble solid formed by trivalent Cr is $Cr(OH)_3$, which, when present, should limit aqueous chromium levels to 50 ppb or less (Palmer and Puls, 1994). The MCL is 100 ppb. Trivalent Cr also forms solid solutions with trivalent iron hydroxides.

ZINC

Zinc (Zn) is mined from sulfide-rich ores and used primarily to make corrosion-resistant coatings (galvanizing) of other metals, and as a component of alloys. Zn is not particularly toxic to humans: people worried about sunburn rub zinc oxide on their noses, and zinc salts are used to solubilize insulin. Zinc exists in aqueous solutions for the most part as Zn^{2+}, though it does form ion pairs with chloride and carbonate as well. Above pH 7 zinc tends to sorb onto a number of minerals. At the same time, zinc hydroxide and carbonate minerals are stable at high pH, where their formation might limit aqueous levels of zinc. At lower pH sorption is less extensive, and zinc is more soluble, and consequently much more mobile. In reducing environments zinc is immobilized as ZnS.

BARIUM

Barium (Ba) is used in oil and gas drilling muds, and consequently shows up in oilfield waste inventories, where it is present as relatively insoluble sulfate minerals. Barium is also used in plastics, pesticides, a variety of oils and fuels, and in a number of medical treatments. Exposure to high levels of barium causes vomiting, diarrhea, and paralysis. Barium exists as Ba^{2+} in sub-alkaline solutions. At pH >9 $BaCO_3$ ion pairs form. In sulfate-rich solutions the $BaSO_4$ ion pair forms. The solids barium sulfate and barium carbonate (respectively barite and witherite) are very insoluble, and grow rapidly. Ba levels in solution are also limited by its appreciable adsorption to clays. The net effect of low solubility (given the right conditions) and non-trivial sorption is that Ba moves very slowly through soils. Although barium is taken up by aquatic organisms and plants, there appears to be little biomagnification in the latter (Smith et al., 1995). The MCL for Ba is 2 ppm.

SILVER, BERYLLIUM, AND NICKEL

Silver (Ag) is widely used in photographic processing and in the manufacture of electrical components (and every now and then to seed clouds). In water silver is generally in the form of Ag^+, though it also forms strong ion pairs with sulfide and chloride. Weaker ion pairing with phosphates, sulfates, and carbonates is also observed. Metallic silver causes permanent, blue-gray staining of the skin. Because of its value, great effort is expended to recycle silver, which in turn decreases the amount which actually shows up in waste streams.

Beryllium (Be) is used in electrical components, in the manufacture of nuclear weapons, in precision instruments, and in ceramics (It is also a component of the gemstone aquamarine). Large amounts of beryllium are also released to the environment by the burning of fossil fuels. Beryllium is a carcinogen, and is known to cause lung damage upon inhalation. Beryllium exists in natural waters as Be^{2+}. It forms oxides and hydroxides having relatively low solubilities Hem (1989). Nickel (Ni) is used to make stainless steel. It is mined from sulfide-bearing ores or from deeply weathered ultra-mafic laterites. The primary aqueous form of nickel is Ni^{2+}, although, like the other divalent cations discussed earlier, its solubility is limited by carbonate and hydroxide precipitation in high pH solutions and sorption to soil components at lower pH. The MCL for nickel is 100 ppb.

NON-METALS

Arsenic is a mining by-product, was once used as an embalming fluid, and, more recently, with lead as a pesticide. Arsenic, like copper and zinc, also shows up in sulfide ore deposits. Arsenic is a common component of chemical warfare agents such as diphenylaminochloroarsine, diphenylchloroarsine, diphenylcyanoarsine, and dichloro(2-chlorovinyl)arsine (respectively, Adamsite, Clark I and II, and Lewisite). Arsenic combines with soil methyl groups easily to form volatile species. In solution arsenic exists in the tri- and pentavalent state. The primary aqueous species are, respectively, $HAsO_2$ and $H_2AsO_4^-$ (low pH) or $HAsO_4^{2-}$ (high pH). Pentavalent arsenic is relatively insoluble and binds to sediments as well. Arsenic is adsorbed through the lungs and intestines and causes a number of cancers in humans. The MCL for arsenic is 50 ppb.

Nitrate, NO_3^-, used in fertilizers, has few natural sinks — mechanisms by which it is rendered unavailable to the biosphere. It sorbs poorly, forms no insoluble minerals, but is converted to N_2 (denitrification; $NO_3^- \Rightarrow NO_2^- \Rightarrow N_2$) during the oxidation of organic matter. When nitrate inputs exceed the denitrification capacity of soils, it accumulates. High nitrate levels have been linked with methemoglobinemia in infants (blue baby syndrome) and drinking water concentrations greater than 44 mg/L (as NO_3^-) are thought to be harmful. The MCL for nitrate is 10 ppm.

RADIOACTIVITY

Radioactive waste is generated from a variety of sources. Uranium was mined and milled for weapons, and then civilian nuclear power, at sites in the United States primarily located in the Four Corners region (northwestern New Mexico, northeastern Arizona, southeastern Utah, and southwestern Colorado). The mill tailings from these processes are radioactive due to the slow decay of ^{238}U and ^{230}Th, and the subsequent decay of their various radioactive daughter products. The most prominent daughter product is ^{222}Rn (Rn = radon) gas, which can cause cancer when inhaled. Mill tailings often have associated with them the hazards derived from the oxidation and/or weathering of pyrite, molybdenum, arsenic, and selenium, as well as the leaching of U. EPA and NRC regulate the cleanup of uranium mill tailings sites, which are volumetrically quite imposing. The Uranium Mill Tailings Remedial Action (UMTRA) program covers 24 inactive processing sites and associated properties. The sites themselves make up nearly a thousand acres and nearly 40,000 cubic tons of waste (Portillo, 1992).

Much high level radioactive waste (HLRW) comes from spent nuclear reactor fuel rods, used for the generation of civilian nuclear power, and from the production of weapon-grade isotopes. Each isotope of an element has a specific total of neutrons and protons in its nucleus. Some isotopes are stable while others decay. HLRW is made up of fission products, and the transuranium elements produced by neutrons captured by the original uranium in the fuel — plutonium, neptunium, americium, and curium. The presence of gamma ray-emitting radionuclides, such as isotopes of cesium, cobalt, and strontium, requires that HLRW be treated remotely, or with dense shielding. Depending on the nature and extent of the fuel reprocessing the net radioactivity of the waste decays to the original radioactivity of the uranium ore body in something like 10,000 to 100,000 years. The amount of HLRW predicted to exist in the United States alone by the year 2000 will be on the order of 10,500 cubic meters (e.g., Faure, 1991). This sounds like a lot, but is equivalent to a football field covered a little over head-high. Ultimate disposal of HLRW in the United States is the responsibility of DOE, as dictated by the Nuclear Waste Policy Act of 1982 (NWPA, 1982). Present plans call for storage of the HLRW produced by utilities in the United States in the proposed repository at Yucca Mountain, in southern Nevada.

Low-level radioactive waste (LLRW) is produced by hospitals, during reactor upkeep, and by scientific laboratories, as well as by the last steps in fuel reprocessing. Much of the radioactivity decays away in a few decades. Defense-generated transuranic wastes are often contaminated with longer-lived iotopes, and are slated to go into the Waste Isolation Pilot Plant (WIPP) in southeastern New Mexico. Commercial and medical LLRW storage is controlled by a number of state compacts set up under the provisions of the Low Level Radioactive Waste Policy Act of 1980. Waste is therefore being stored in a number of locations. In the past much was stored at sites in Beatty, Nevada; Barnwell, South Carolina; and Richland, Washington. In most cases the waste has been stored in trenches in a variety of containers having projected lifetimes of a few decades at best. The U.S. DOE controls much of the high and low-level radioactive waste, and has responsibility for a number of sites contaminated by radioisotopes, as well as organic contaminants, during the cold war arms race. DOE facilities cover roughly 2800 square miles (Riley et al., 1992), and the primary sites are in New York (Brookhaven National Laboratory), Ohio (Fernald, Mound, and the Portsmouth Gaseous Diffusion Plant), South Carolina (Savannah River), Florida (Pinellas), Tennessee (Oak Ridge National Laboratory), Missouri (Kansas City), Illinois (Argonne National Laboratory), Colorado (Rocky Flats), Texas (Pantex), New Mexico (Sandia and Los Alamos National Laboratories), Idaho (Idaho National Engineering Laboratory), Nevada (The Nevada Test Site), California (Livermore and Sandia-Livermore), and Washington (The Hanford Site). The largest fraction of contamination is at three facilities; Hanford, Savannah River, and Oak Ridge. Contamination of soils and groundwaters by metals, radionuclides, organic solvents, fuel hydrocarbons, and other miscellaneous chemicals came about due to the enrichment of uranium, the production of plutonium and tritium, weapons testing, and nuclear power research. Some of the waste resides in large tanks which await final treatment and disposal. Soil and groundwater contamination, though, has resulted from the historic disposal of wastewater solutions in ponds, cribs, and basins, as well as from leaks. Often radionuclides reside with toxic metals and/or hazardous organics. Because it is difficult to predict the chemical behavior of these mixed wastes in tanks, let alone complex soils, the technical challenge of cleanup is formidable. At the same time, because DOE's facilities are spread throughout the country contamination is found in a wide variety of soil types, adding to the uncertainty in predicting long-term transport. The cleanup cost for this set of sites has been euphemistically termed "The Cold War Mortgage", and has been estimated to run in the hundreds of billions of dollars. Figures 3.3 and 3.4 point out the various radionuclides and organic solvents about which DOE is particularly concerned, and give an idea of the magnitude of the cleanup problem.

Natural attenuation processes were first recognized in connection with our understanding of radionuclides. A contaminant's time on the environmental stage is set by its half-life, and comparing contaminant half-lives allows them to be pigeonholed in terms of persistence and health impacts. This is true for radionuclides, as well as non-radionuclide organics which are biodegraded. Radionuclides with long half-lives are long-term concerns, but have correspondingly lower specific activity. Those with short half-lives are less of a problem over the long haul, but are more of a health threat over the short term. Health effects of radiation come from the energy and/or particle release that occurs upon decay. Radionuclides which decay rapidly thus carry a proportionally higher potential to impact human health, all other things being equal. Because different elements are taken up by different parts of the body, and different isotopes have different decay schemes, health effects are isotope specific. Extensive tables exist from which the particular scaling factors among dose, organ, and half-life can be assessed (e.g., Eckermann et al., 1988).

Sorption and ion exchange are important natural sinks for many radionuclides (e.g., U, Pu, Am, Ra) that form positively charged species in solution and therefore

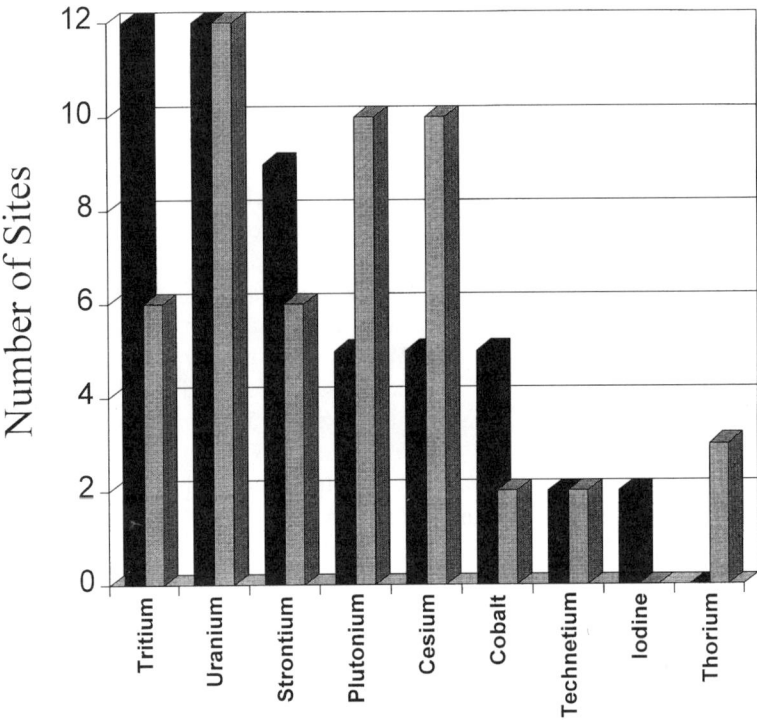

FIGURE 3.3 Number of DOE sites contaminated with radionuclides. Black bar, groundwater; gray bar, soils. (From Riley et al., 1992).

stick to most soils, which are, by and large, negatively charged. At the same time some radionuclides such as thorium, Pu^{4+}, and americium, are very insoluble in most soils. Radionuclides in soils and groundwaters might be regarded as naturally attenuated if they are retained in, or on, mineral phases for ten half-lives or more. Obviously, there are complications if the decay products of an isotope are radioactive as well. For example, ^{238}U decays to produce ^{226}Ra (Ra = radium), which has a half-life of 1620 years. ^{226}Ra decays to ^{222}Rn which has a half-life of roughly 4 days and, once ingested, decays to other, longer-lived radioactive isotopes, which, along with the radon can cause lung cancer. According to the ten half-life rule-of-thumb, a little more than a month is needed for the health effects of radon to be minimized. Because radon is a gas that is quite mobile, 40 days is sufficient for its widespread movement out of ^{238}U-bearing soils (and mine tailings) and into houses.

Some radioactive elements, such as I and Tc, sorb very weakly to most soil minerals, and tend to desorb readily. ^{131}I, with a half-life of 8 days, was one of the volatiles released from the nuclear reactor accident at Chernobyl in the U.S.S.R in April, 1986. ^{137}Cs made up a great deal of the radioactivity released from Chernobyl as well. Cs sorbs weakly to mineral edges, but over time becomes more strongly bound to clay inner layers (Comans et al., 1991). Very few natural sinks exist for tritium, ^{3}H, produced by thermonuclear bomb tests and in nuclear reactors; however

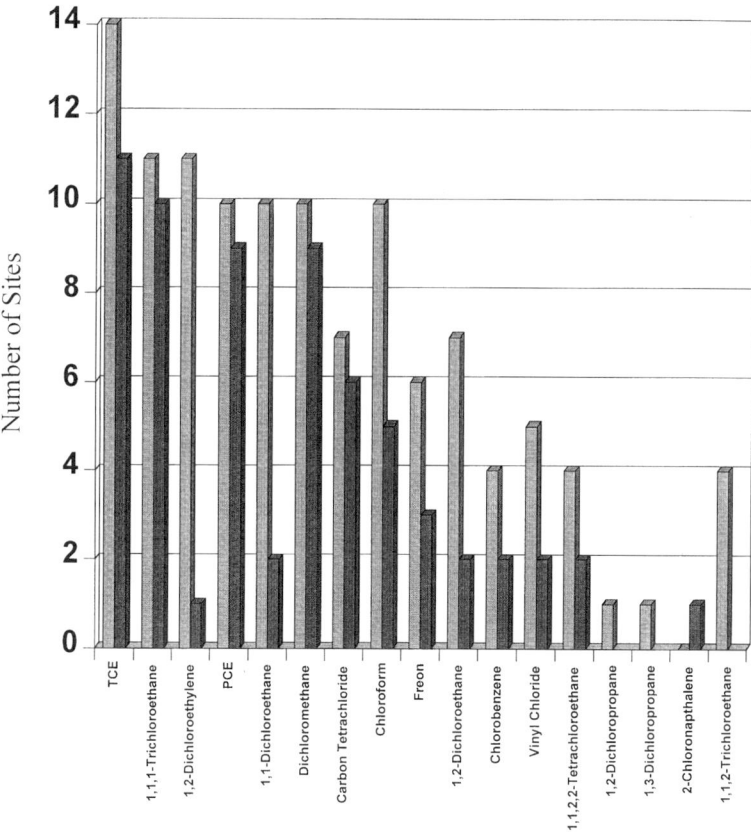

FIGURE 3.4 Number of DOE sites contaminated with chlorinated hydrocarbons. Gray bar, groundwater; dark gray, soils. (From Riley et al., 1992).

its half-life for elimination from the body is only 3 to 4 weeks, and its half-life is ~12 years, so its long-term environmental consequences are minimal, except where acute exposure is an issue.

ORGANICS

Stumm and Morgan (1981) pointed out that the number of synthesized organic compounds in the early 1980s approached 2 million, with an addition of ~250,000 new ones each year. Although only 300 to 500 of the latter made it into industrial production, between 100 and 200 million tons of synthetic organic chemicals were being produced per year, with an estimated 30% of the latter ending up in the environment. The amount of synthetic organic matter being produced now exceeds 300 million tons per year, and the total amount of synthetic organic chemicals entering the environment approaches 15% of the carbon flux arising from natural, primary production (Egli and Bally, 1996). Some of these organic compounds are carcinogenic, others have non-carcinogenic health effects as well.

PETROLEUM HYDROCARBONS

There are some general rules of thumb used to predict the behavior of organic compounds in soils that will be covered quantitatively in Chapters 6 and 7. In general, the higher the carbon number of a given organic molecule, the more pronounced is its tendency to adhere to organic matter in soils, the lower is its solubility in water, and the smaller is its potential to turn to vapor. Many of the lighter components of fuels are rapidly biodegraded in the subsurface, leaving the more resistant compounds behind. The net result is that the chemical makeup of petroleum-based contaminants changes over time in soils and reflects the net of volatilization, biodegradation, and sorption.

The average gas station has at least two underground fuel tanks; most of the older ones leaked. In 1986 the U.S. EPA estimated that 35% of the nation's underground fuel tanks were leaking (U.S. Environmental Protection Agency, 1986). In gasoline the hydrocarbon components are primarily C4 to C12; that is they possess between 4 and 12 carbon atoms. The lighter of these (those with lower carbon numbers) volatilize readily into soils. Aliphatic hydrocarbons are linear (or branched) chains of carbon atoms joined by single (alkanes), double (alkenes), or triple (alkynes) bonds. The primary aromatic components (those built up from a benzene ring) are benzene, toluene, ethylbenzene and xylene (collectively termed BTEX — see Figure 3.5), and make up over 50% of the water-soluble component of unleaded gasoline (Coleman et al., 1984), and 10 to 40% of the total. Some polycyclic aromatic hydrocarbons (PAHs — two or more linked benzene rings) are also present. Jet fuel is similar to kerosene in that it is made up of hydrocarbons in the C11 to C13 range. Diesel fuels fall in the C10 to C20 range. Fuel oils, used for home or industrial heating, or as blending agents, generally possess higher levels of alkylated phenanthrenes and napthalenes, and a larger component of nitrogen, sulfur, or oxygen. The MCLs for benzene, toluene, ethylbenzene, and xylenes are, respectively, 5, 1000, 700, and 10 ppb.

NITROAROMATICS

Nitroaromatics are benzene rings having one or more NO_2 (nitro) groups attached (see Figure 3.6). The nitroaromatics that are found in soils and groundwaters are almost invariably man-made, and are widely used for pesticides, dyes, and as explosives. A number of pathways leading to breakdown of nitroaromatic compounds by microorganisms have been demonstrated (Bradley and Chapelle, 1995; Costa et al., 1996; Shelley et al., 1996; Spain, 1995). One of the better-studied breakdown paths is that of 2,4,6-trinitrotoluene (TNT). TNT contamination exists at a number of munitions sites. TNT consists of a toluene whose methyl group ($-CH_2-$) separates two nitro groups and which is directly opposite (across the benzene ring) a third (see Figure 3.6). TNT sorbs effectively to many soil minerals. TNT also is observed to biodegrade under the right conditions.

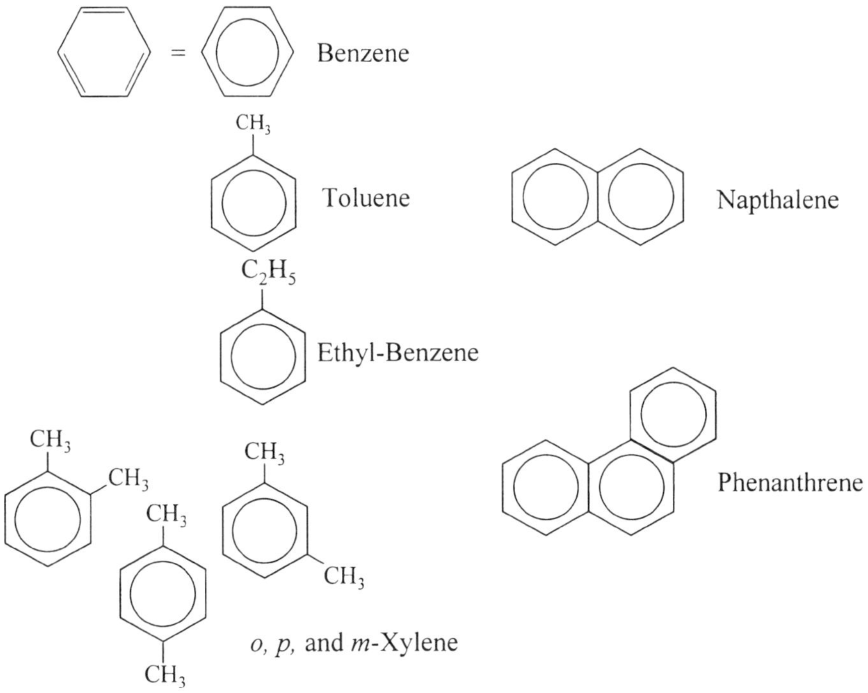

Sources of Hazardous Waste

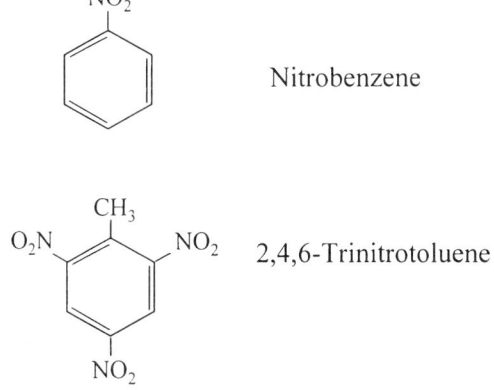

FIGURE 3.5 Benzene, toluene, ethylbenzene, xylenes, phenanthrene, and napthalene.

FIGURE 3.6 Nitrobenzene and 2,4,6-Trinitrotoluene (TNT).

HALOGENATED HYDROCARBONS

Halogenated hydrocarbons (carbon groups with attached halogens — chlorine, bromine, and fluorine) are ubiquitous industrial products, and are used as aerosol propellants, foaming agents, in dry cleaning operations, metal cutting, and occasionally as pesticides. They can also be produced in trace amounts by natural processes such as the burning of vegetation. The EPA list of priority pollutants is dominated by chlorinated organic compounds such as: trichloroethene (TCE, a solvent used in dry cleaning and metal degreasing; also a degradation product of tetrachloroethene; sometimes written as trichloroethylene), methyl bromide (a pesticide), tetrachloroethene (PCE, a solvent used in dry cleaning and metal cutting; also written as perchloroethene, tetrachloroethylene, or perchloroethylene), and vinyl chloride (used to manufacture PVC; also a breakdown product of trichloroethylene) which is highly toxic and causes angiosarcoma, a liver disease. Other halogenated hydrocarbons found in hazardous waste include: *o*-dichlorobenzene (a solvent, fumigant, and insecticide) and chloroform (a solvent and insecticide). The MCL for TCE and PCE is 5 ppb.

Polychlorinated biphenyls (PCBs) are a particularly important class of chlorinated aromatic compounds. They have a low vapor pressure, and, consequently, don't volatilize readily. At the same time they are excellent electrical insulators and were used in the making of transformers and capacitors. PCB's don't biodegrade rapidly. They tend to accumulate in the food chain, and collect in fatty tissue in humans. Exposure to PCBs leads to reproductive disorders and tumors. Consequently, the current use of PCBs is severely regulated. The MCL for PCB's is 0.5 ppb.

Biodegradation and volatilization are two important processes which remove halogenated organics from soils. Chlorinated hydrocarbons are degraded by soil microbes, though generally less rapidly than non-chlorinated hydrocarbons, such as fuel components (see below). Another removal mechanism, or sink, for long-lived organic contaminants which organisms leave behind appears to be sequestration inside mineral particles. Alexander (1995) cites the following evidence:

1. The bioavailability of some long-lived organic chemicals decreases over time in the field.
2. Freshly added chemicals appear to be taken up from solutions wherein aged versions of the same compound are not.
3. Some compounds become increasingly resistant to extraction with time.
4. Sorption and desorption are found to take long periods of time to reach equilibrium.

The net effect of contaminant movement into dead-end pores is that the available toxicity (i.e., health risk) of the contaminant disappears faster than the chemical itself.

PETROLEUM PRODUCTION WASTES

Onshore drilling operations generate between 100 and 150 m^3 per well of oil-based waste (Chaineau et al., 1995), consisting of petroleum-contaminated drilling fluid (+ additives) and rock. Biodegradation of the organic compounds is commonly observed, and there is a general reliance on land-farming for disposal. While the lighter components are rapidly biodegraded in many soils, many of the heavier compounds are resistant (e.g., Bossert and Bartha, 1984; Oudot et al., 1989). In addition to the organic contamination it should be noted that traces of radioactive isotopes of uranium, potassium, radium, and thorium are often found in petroleum reservoir rocks and wastes. The latter are referred to in the industry as Naturally Occurring Radioactive Material (NORM). Radioactivity shows up in scale on pipes, in films formed during production or processing, and in sludges in pipelines and processing plants (Gray, 1990). NORM contains uranium, thorium, radium, and their decay products. The primary health concern of NORM is the radon produced as a decay product.

LANDFILL LEACHATES/ACID MINE DRAINAGE

There were nearly 13,000 landfills in the United States in 1983, and nearly 2400 open dumps (Peterson, 1983). Because of the heterogeneous nature of garbage, landfills contain a wide spectrum of metals and organic contaminants. Landfills are designed to isolate wastes from the biosphere. Most landfills have low-permeability liners (e.g., clay) on the bottom and, in many cases, low permeability covers on top, to prevent the seepage of rainwater into the fill. Leachate forms due to the presence of fluids in the waste and/or the seepage of rainwater. Many modern landfills have collection pipes to remove leachate, and thus prevent transport of landfill toxicity to drinking water. Older landfills were often sited more on the basis of convenience, and included gullys, arroyos, ditches, and the like, where they are prone to leaching.

Acid mine drainage forms when water and atmospheric oxygen come into contact with metal sulfide minerals left behind by mining activities. The resulting solutions may be rich in heavy metals and sulfate, and have very low pH's. Eventually, acid solutions produced by natural processes are ultimately neutralized by reaction with base-containing metal carbonate and silicate minerals. Limestones (Ca and Mg carbonate minerals) dissolve quite rapidly and are particularly effective at neutralizing acid mine drainage when the two come into contact. Neutralization of acid mine drainage causes iron (hydr)oxides to precipitate and, in the process, to sorb and often remove a fraction of the heavy metals from solution; however, the distance travelled by the metals before this occurs may be long. For an extensive review of the environmental geochemistry of acid mine drainage, see Alpers and Blowes (1994).

BACKGROUND READING

Fetter C. W. (1993) *Contaminant Hydrogeology.* Macmillan, New York.
Hem J. D. (1989) Study and interpretation of the chemical characteristics of natural water. United States Geol. Surv. Water Supply Paper 2254.
Manahan S. E. (1991) *Environmental Chemistry.* Lewis Publishers, Boca Raton, FL.

4 Groundwater Flow

INTRODUCTION

Movement of hazardous substances in soils and groundwaters can be very fast, on the order of meters per day, or very slow, with only small distances (meters or less) covered over time spans approaching the geologic. The primary transporting agents are soil fluids; however, in cases where the contaminants volatilize, vapor transport can be significant as well. Obvious examples of contaminants which partition into the vapor phase are volatile organics such as BTEX compounds, and methylated metals such as Hg and As. Physical processes which can lower contaminant levels in soils are dilution and dispersion, the "smearing out" of plumes through random fluid motions and chemical diffusion. Fluid velocities are also important as they affect both dilution and dispersion. More importantly, absolute fluid velocities often limit the actual potential impact of a plume on public health, as well as the likelihood that remediation schemes such as pump-and-treat will achieve their goals. Plumes that move on the order of centimeters per year or less are probably not pressing health threats, as long as drinking water is not being drawn from them, or if drinking water is being treated at the wellhead.

DARCY'S LAW

In the rare absence of chemical degradation mechanisms and/or volatilization potential, public health risks often depend on the physical and chemical factors which control fluid flow rates. The most important of these factors is hydraulic conductivity, K_h (meters/day), followed by effective porosity, θ, and the hydraulic gradient $\Delta h/\Delta l$. The equation relating all of these is a modification of Darcy's law:

$$V = \frac{K_h}{\theta} \frac{\Delta h}{\Delta l}$$

V is the velocity (meters/day) of fluid movement (seepage velocity) through pore spaces in the soil. $\Delta h/\Delta l$ is simply the distance, Δh, the water table drops over the distance the water travels, Δl. The negative sign is an easily remembered convention pointing to the fact that water flows down hill, in the sense that the water table is the hill. The topography of the water table determines the general direction of fluid flow, and, potentially, plume migration. It is therefore critical in site assessments that the topography of the water table be accurately determined. Darcy's law states that, all other things being equal, particularly K_h: (1) the steeper is the hill (i.e., the greater is $\Delta h/\Delta l$) the faster the water will move, and (2) the more confined the path, i.e., the smaller is θ, the faster water will flow, as measured in the pore.

Before going further, an important distinction must be made between the portion of soils wherein the pores, or open spaces, are almost fully occupied by water, and the shallower portions where the pores are filled mostly with air. The first is referred to as the saturated zone, the latter is termed the unsaturated, or vadose, zone. A "capillary fringe" separates the two where water exists as a thin film in intermittent contact with the soil matrix. Darcy's law, as written above, describes fluid flow in the saturated zone. The relation can be used though, with a few modifications, to describe fluid, and dissolved contaminant flow in the vadose zone as well (see below). In real life it all depends on where the contaminant is. In arid climates, such as the American southwest, the saturated zone is often quite deep, sometimes hundreds of feet below the land surface. In eastern North America or continental Europe and near any permanent stream or river, the saturated zone often lies within a few feet of the land surface. Here we will deal with the limiting case of fluid movement in the saturated zone, leaving the special cases of unsaturated flow and mixed-phase flow for the end of the chapter.

Returning to Darcy's equation, hydraulic gradients in soils vary depending on the hydrologic state of the aquifer. An aquifer is defined as a rock body containing water. In most soils, the capillary fringe marks the top of the saturated zone (as opposed to a confining layer of less permeable rock), and the configuration is termed an "unconfined aquifer". The water table often grossly mimics topography. Therefore, the grade of the water roughly approximates that of the land surface, allowing some limits to be set on the range of gradients to use in Darcy's law to approximate groundwater, and dissolved contaminant, velocities. On the low end, for flat water tables, $\Delta h/\Delta l$ can approach zero, in which case there is no movement of fluid. Note though that contaminant movement can still occur through dispersion (see below). On the high end, in very steeply sloping, unconfined aquifers $\Delta h/\Delta l$ can exceed 0.1. In any case, the gradient is a number between 0 and 1, and very easy to measure in the field by determining water level differences relative to a common base level.

This leaves porosity (θ) and hydraulic conductivity (K_h), two variables which depend on the physical nature of the rocks and soils themselves. A number of attempts have been made to relate the latter, which is relatively hard to measure, with the former, which is somewhat easier. For a review of the various empirical relations proposed between the two, see Lerman (1985). Porosities vary by roughly a factor of 5 to 10 in soils. High values are on the order of 0.5. Low values in tightly packed rocks are on the order of 0.1 or less. The critical uncertainty in fluid transport is often K_h, which can vary by several orders of magnitude. In gravelly soils having large, connected pore spaces K_h can be as high as meters per day. Sandy soils have somewhat lower conductivities and K_h is in the range of mm's to tens of cm's per day. In soils having large quantities of clays, hydraulic conductivities dwindle due to the relative lack of channels, and hydraulic conductivities can be less than cm's per year. Obviously, the value of K_h has a lot to do with whether a water-borne contaminant has the potential to be a problem or not. Hazardous substances dissolved in a clay-rich aquifer having a hydraulic conductivity of 1 mm per day and moving at the same speed as the water will travel only 18 meters in a thousand years, given typical values of θ and $\Delta h/\Delta l$ of, respectively, 0.2 and 0.01. On the other hand water-borne toxins in a gravel aquifer with the same values of θ and $\Delta h/\Delta l$, but a K_h of 1

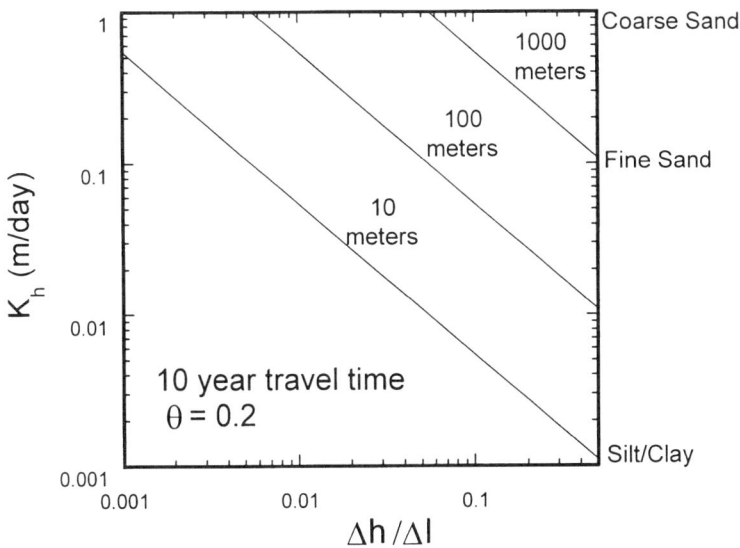

FIGURE 4.1 Ten-year travel time for groundwater calculated from Equation 4.1 for a variety of hydraulic conductivities and gradients assuming porosity equals 20%.

meter per day, should have a thousand-fold greater rate of movement of 18 kilometers per thousand years. Figure 4.1 uses a porosity of 0.2, a combination of hydraulic conductivities and gradients, and Darcy's law to calculate the distance a fluid can travel in ten years.

All this being said, a few caveats are in order. Even in soils having small hydraulic conductivities (as determined on relatively small and homogeneous lab samples) the existence of even minor cracks and fractures can increase the hydraulic conductivity by several orders of magnitude and provides a much faster path for contaminant transport. This raises the point that soils are complex, heterogeneous mixtures of minerals, and that models which attempt to describe the physical and chemical characteristics of soils are, at best, very rough approximations. Because of this heterogeneity it is often impossible to predict exactly where contaminants will move in the subsurface. This makes it incredibly hard to come up with the optimal cleanup solution for a given site. For the purposes of assessing natural attenuation numerical precision is not so important, as order-of-magnitude assessments of soil hydrology and reactivity often provide useful answers to questions regarding the long-term behavior of pollutants in a particular setting.

In addition to heterogeneities caused by fractures, it should be pointed out that hydraulic conductivities in otherwise uniform-appearing soil vary significantly over very short distances. Also, because of the directional makeup of rock fabrics, hydraulic conductivities often vary depending on the direction the water is moving. In general hydraulic conductivities for vertical fluid flow are smaller than horizontal hydraulic conductivities, typically by at least a factor of 2 or 3. Hydraulic conductivity may be measured in the field by a variety of pump tests involving the measurement of an aquifers response (local change in water level) to water withdrawal,

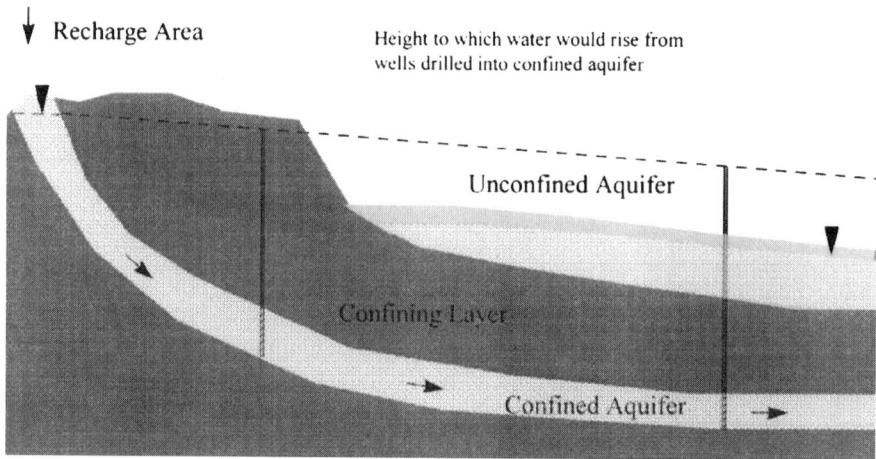

FIGURE 4.2 Schematic of confined and unconfined aquifers and their respective potentiometric surfaces.

or injection. For a summary of the various methods see Domenico and Schwartz (1990).

The same approach to fluid movement in the saturated zone of unconfined aquifers can be applied with some modification to set limits on fluid movement at much greater depth, specifically in contaminated aquifers overlain by confining layers of rock. Municipal drinking water in North America comes largely from surface water and sand, gravel or limestone aquifers which are tens to hundreds of feet below the water table. In such aquifers the gradients used in Darcy's law are not those of the water table, but are rather, specific to the aquifer itself. In Darcy's law, h represents the height of the water column measured by a piezometer placed in the aquifer. The 3-dimensional distribution of heads for an aquifer is the potentiometric surface, and is analogous to the water table case for unconfined aquifers (see Figure 4.2), but in this case the surface can be thought of as the water levels measured by an array of cased wells that only open into the aquifer of interest. In confined aquifers this is a dynamic surface defined by a host of factors including: the water balance in the aquifer, the configuration of the underlying and overlying strata, and the regional hydraulic conductivity pattern (to name the most important). The same caveats covering hydraulic conductivities mentioned earlier apply also for confined aquifers.

UNSATURATED FLOW, DNAPLS AND LNAPLS

In the unsaturated zone water exists discontinuously (see Figure 4.3) and moves in response to tension, or equivalently, matrix suction. Matrix suction is highest where the moisture content is lowest and vice versa, though this depends on the actual surface area because matrix suction arises due to the capillary forces caused by adhesion of water to soil particles. With a level water table the gradient used in Darcy's law is the gradient in matrix suction, $-\Delta\Psi/\Delta l$, where Ψ is matrix suction

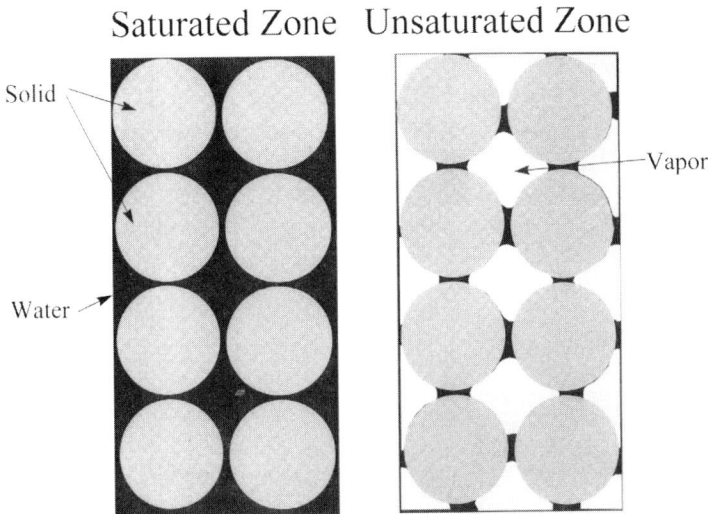

FIGURE 4.3 Saturated and unsaturated porous media.

(in meters). Water therefore flows from moist soils to soils with lower moisture content. The capillary fringe is a direct result of this movement. Values of K_h in the vadose zone also vary with moisture content, being greatest when the soil is saturated. Therefore, to use Darcy's law in the unsaturated zone requires that K_h be expressed as a function of saturated porosity. A useful overview of how this is done in particular, and how unsaturated transport works in general, can be found in Wierenga (1995).

Multiphase flow is a critical environmental concern as a large variety of organic contaminants form separate liquids in soils. The latter are termed non-aqueous phase liquids (NAPLs) and are either more dense (DNAPLs) or less dense (LNAPLs) than water. NAPLs form when the forces for self-attraction between the NAPL molecules exceed those between NAPL and soil solutions. LNAPLs can, if present in sufficient quantity, pool at the capillary fringe, sometimes collapsing it in the process. Flow of LNAPLs follows the gradient of the water table. DNAPLs are particularly troublesome as their sinking through the saturated zone is relatively unhindered by the gradient of the water table. This makes it very difficult, and often impossible, to predict their motion in the subsurface.

Contaminant fluids which are immiscible in water are still capable of continuously dissolving tangible fractions of their components into solution during their whole transit through a given soil. The actual amount of dissolved contaminant depends on a variety of chemical factors to be covered in the following chapter. NAPLs in general, and DNAPLs in particular, also can potentially coat (or equivalently, "wet") mineral surfaces. A layer of water coats mineral surfaces in the saturated zone, and to a large extent, minerals in the vadose zone as well. Many NAPLs have ionized (charged) functional groups that interact with the broken bonds on the mineral surface underlying the water layers. If this interaction is strong enough to expel the intervening water molecules the ionized groups of the NAPL are able to adhere to the mineral surface. The chemical details of this will be discussed later,

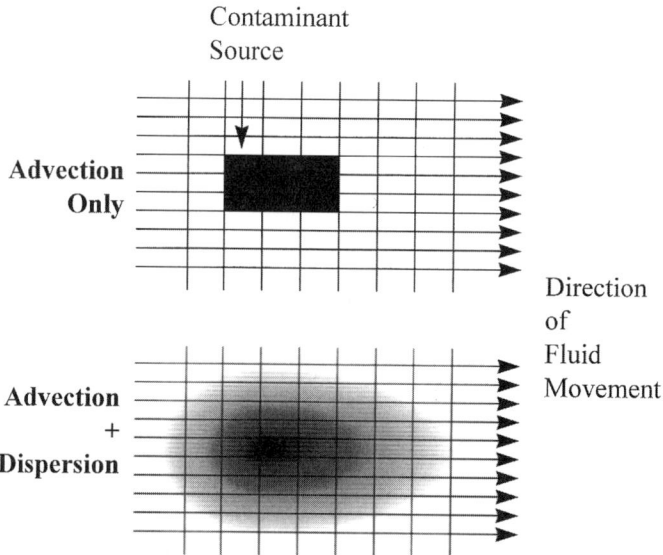

FIGURE 4.4 Plume behavior for pure advection and advection + dispersion.

but it should be pointed out that in a physical sense adhesion of NAPLs to mineral surfaces makes their removal by standard pump-and-treat methods quite difficult because more volumes of water are required to dissolve and/or detach NAPLs from the aquifer material.

DISPERSION

Contaminants move in the subsurface by advection and by dispersion. Advection is most simply thought of as the case where a dissolved chemical moves at the same speed as the carrier fluid. Advective velocities are nothing more than the fluid velocities calculated from Darcy's law. Dispersion is the "smearing out" of a plume by a combination of chemical and physical factors. Fresh recharge can cause mixing and physical dispersion, as can seasonal fluctuations in the level of the water table. Dispersion causes a net drop in contaminant levels at any given point in a plume at the expense of increasing the total volume of water contaminated (see Figure 4.4). When contaminant levels in a plume are at, or near, the regulatory limit dispersion can in theory "remediate" a site by dropping levels to the required target. To examine dispersion more closely it is useful to split it into its component parts, chemical diffusion and true dispersion. Chemical diffusion of a contaminant i is described by Fick's first law:

$$F_i = -D_{d,i} \, \Delta C_i / \Delta l$$

F, in this case, is the flux of contaminant i in grams per m² per second. $D_{d,i}$ is the diffusion coefficient, which is specific to a given chemical i in a given solution. $D_{d,i}$

is around 10^{-9} meter^2/second for many molecules. $\Delta C_i/\Delta l$ is the concentration gradient which drives the diffusive flux; i.e., ΔC_i is the difference in concentration of the contaminant over a distance, Δl, the contaminant is to move. The negative sign is a convention to emphasize that the contaminant molecules diffuse from high to low concentrations. Because mineral grains get in the way of diffusing molecules, when modeling diffusion in porous media the diffusion coefficient in the equation above is generally multiplied by a factor reflecting the tortuosity of the flow path. Tortuosity is simply equal to L_e/L; where L_e is the length of the path a molecule travels between two points and L is the straight line destance between the same two points. Diffusion distances are generally on the order of cm's/yr or less, which points up the fact that chemical diffusion generally moves contaminants too slowly in soils and groundwaters to achieve a substantial redistribution over anything much short of geologic time scales.

Examination of fluids moving through the subsurface at the molecular level would reveal water moving at a spectrum of velocities differing significantly from the mean calculated by Darcy's law. This distribution makes itself felt on dissolved molecules most notably by moving them at varying velocities. At the same time there is a lateral spreading of molecules at the microscopic level due to differences in flow path geometry (see Figure 4.5). The net effect is observed on the macroscopic level as a "smearing" of contaminant plumes (see Figure 4.6). While the origins of mechanical dispersion are grasped readily, it is impossible to predict in detail the nature and extent of dispersion in natural systems. Longitudinal dispersion (dispersion in direction of flow) increases with the length of the flow path. Differences in hydraulic conductivity are the primary cause of dispersion in the field. An extensive discussion of dispersion and its effects on transport modeling can be found in Fetter (1992). For simple groundwater modeling, a number of empirical relations exist which allow dispersion to be roughly estimated (see Newell et al., 1996). Specifically, dispersion in one dimension, D_x is written as:

$$D_x = \alpha_x v_x + D^*$$

α_x is the longitudinal dispersivity; v_x is the average linear groundwater velocity; and D^* is the effective molecular diffusion coefficient. α_x has been estimated as a function of L_p, the plume length, by a number of authors: $\alpha_x = 0.1 L_p$ (Pickens and Grisak, 1981); $\alpha_x = 3.28 \cdot 0.83 \cdot [\log(L_p/3.28)]^{2.414}$ (Xu and Eckstein, 1995). Transverse dispersivity, α_y, has been estimated to fall between $0.1\alpha_y$ and $0.33\alpha_y$ (see Newell et al., 1996). Vertical dispersivity, α_z, for most cases is quite small, and has been estimated to be between $0.025\alpha_z$ and $0.1\alpha_z$ (see Newell et al., 1996).

COLLOIDAL TRANSPORT

Colloids are small water-borne particles that are between a billionth and a millionth of a meter in length. Colloids can be mineral fragments such as clays, organic macromolecules, microorganisms (such as bacteria and viruses), or biological debris. Colloids are removed from groundwater and soil solutions when they aggregate and

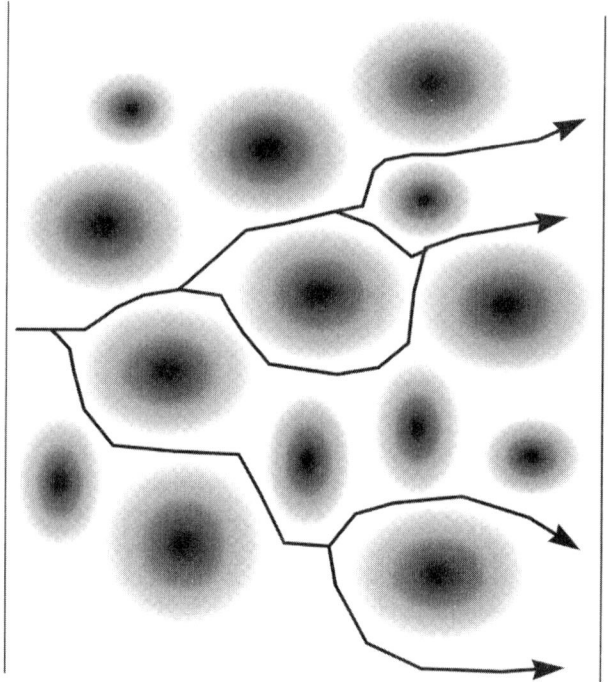

FIGURE 4.5 Lateral dispersion due to molecular movement. (After Fetter, 1992).

settle, or when they become stuck through chemical bonds or physical trapping between mineral surfaces. Colloids may remain suspended for extended periods of time, and have been observed to travel hundreds of meters through aquifers. Because colloids can sorb contaminant molecules, or themselves be made of contaminant molecules (e.g., a number of actinides are known to form these "intrinsic" colloids), colloidal movement constitutes an additional pathway for contaminant transport. Figure 4.7 shows a size distribution of water-borne and other soil particles. Lead, plutonium, and americium, which dissolve only sparingly into the aqueous phase, are seen to be occasionally transported by colloids (McCarthy and Zachara, 1989; Wang et al., 1995). The same behavior might be expected for otherwise strongly sorbing organic contaminants when organic colloids are abundant.

TRACERS

The accumulated uncertainties in hydrologic models are large. In particular, the heterogeneties in hydraulic conductivity, potential presence of "fast paths", and the general unpredictability of dispersion prevent hydrologic models from ever predicting subsurface reality in more than a gross sense. Nevertheless, advection and dispersion often must be known rather precisely for natural attenuation to be demonstrated. If advection and dispersion are unknown it is very difficult to show unambiguously that contaminant degradation is occurring, or for that matter, how important it is.

Groundwater Flow

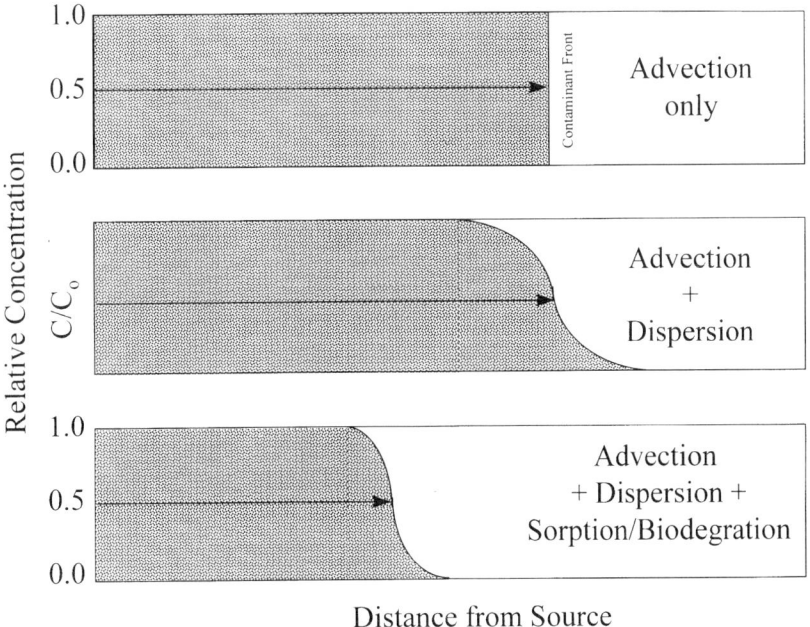

FIGURE 4.6 Relative effects of advection, dispersion, and biodegradation on plume breakthrough from a continuous source.

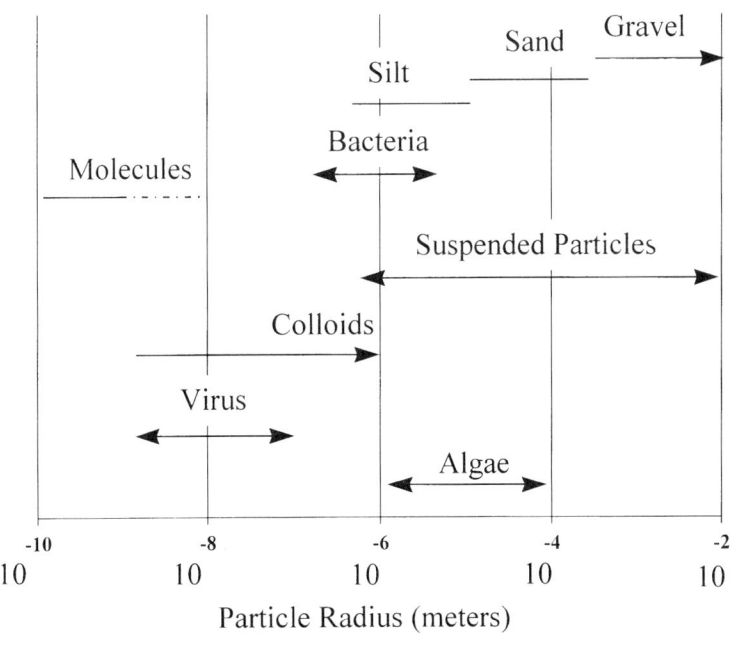

FIGURE 4.7 Particle sizes in soils.

One way to get around the incomplete picture of advection and dispersion is to instead rely on non-reactive chemical tracers to determine the flow field and the magnitude of dispersion. A tracer is a chemical which specifically does not adsorb, become biodegraded, form insoluble minerals, or volatilize appreciably. In other words, a tracer is inert in a biogeochemical sense and its movement in the subsurface tracks the baseline transport of contaminants due solely to physical processes (advection and dispersion). Consequently, the outline of a tracer plume defines how far a contaminant would have gone in the absence of chemical reactions with soil solids. If contaminant and tracer plumes are identical, no reaction has occurred. On the other hand, a clear lag between the tracer and the contaminant plume unambiguously points to one, or a series, of contaminant sinks in the subsurface.

Chloride and bromide, and detergents are commonly used as tracers. There is a difference between performing a tracer test using a chemical injected at a wellhead to work out the flow field in an aquifer, and using tracer behavior to separate out dilution and dispersion from sorption and biodegradation. In the common case where a plume has existed for a number of years and the desire is to identify contaminant sinks, a non-reactive component of the original source must first be identified. MTBE, methyl tert butyl ether, is an inert compound, present in most leaking underground fuel tanks, which might be used as a tracer (Rifai et al., 1995a). This contrasts to the situation where one (or more) injection wells have been constructed, and tracers, such as SF_6, are injected and tracked to unravel subsurface behavior.

ENDNOTE

Fluid transport at depth is controlled by hydraulic conductivity, and, to a lesser extent, physical dispersion. Except for the trivial case where dilution and dispersion can "remediate" a site, physical hydrology is important for two reasons: (1) It gives clues to the direction and velocity of fluid and contaminant movement, and (2) It determines the amount of time from source to receptor that contaminants are vulnerable to degradation mechanisms. In other words, physical hydrology provides the distance component of the *persistence to distance* assessment of potential contaminant impact. The persistence portion is determined by chemical and biological processes. The clearest macroscopic manifestation of contaminant degradation is a shrinking or stable (non-expanding) plume. A shrinking plume means that contaminant mass is being decreased by some process. If the source is not removed, this is equivalent to the net of degradation mechanisms being greater than the flux from the contaminant source. A stable plume, in the absence of a decrease in loading from a source, indicates that the contaminant degradation rate equals the contaminant flux from the source. Even expanding plumes can point to natural attenuation of contaminants; namely, if the plume movement predicted by a hydrologic model overestimates the observed plume movement, there is a chance that the difference, the retardation in plume movement (and presumably reduction in contaminant mass) is due to degradation along the contaminant flow path.

BACKGROUND READING

Domenico P. A. and Schwartz F. S. (1990) *Physical and Chemical Hydrogeology.* Wiley-Interscience, New York.

Dragun J. (1988) *The Soil Chemistry of Hazardous Materials,* Hazardous Materials Control Research Institute, Silver Spring, MD.

Fetter R. W. (1993) *Contaminant Hydrogeology.* Wiley-Interscience, New York.

5 Chemical Attenuation

INTRODUCTION

All soil contaminants face a trip of varying distance and difficulty between their starting point and the point where they might constitute a health hazard to humans, or where they violate a regulatory limit. The latter point is often the wellhead tapping a shallow aquifer for drinking water, or a stream, or a river. By no means is a deleterious impact on health inevitable. The greater the travel distance, the greater is the likelihood that the contaminant will be diluted and dispersed to the point where it poses no health risk. Nevertheless, in the absence of chemical interactions with soil minerals, biota and solution, each contaminant molecule might be expected to reach a downstream receptor, given enough time. While dilution and dispersion are the only physical processes able to lower contaminant levels, there are a host of chemical processes capable of lowering contaminant levels below regulatory levels. Because of the efficacy of chemical retardation, a fair bit of elaboration on its origins is needed. The focus here will be on inorganic processes. Microbiological pathways to natural attenuation will be explored in the chapter which follows.

In order to roughly classify the fate and transport characteristics of contaminants in groundwaters it is necessary to define a number of terms and outline the basic chemical controls on dissolved contaminant levels. The bulk of contaminant reactivity can be treated as one of a combination of seven types of chemical reactions: ion pairing, gas phase dissolution, mineral growth, mineral dissolution, electron transfer, adsorption, and radioactive decay. All of these reactions besides radioactive decay can be reversible. Reversible reactions are easiest understood by first defining their opposite number, irreversible reactions. In irreversible reactions, reactants are turned into products at a rate which doesn't depend on how much product has accumulated. An ice cube melting in warm sunlight does so at a rate that doesn't depend on how much water accumulates. In other words, the reaction: ice $\Rightarrow$ water can't be reversed by changing the ratio of ice to water. Hence, the reaction is termed irreversible. In Earth surface environments radioactive decay is irreversible by the same reasoning (Another example of an irreversible reaction is the dissolution (weathering) of a mineral formed at high temperatures (e.g., a quartz or feldspar) into dilute, low-temperature waters). Reversible reactions can be turned in the opposite direction if the product-to-reactant ratio is increased (or if temperature or pressure are changed). If a gas containing CO_2 is bubbled through water a fraction of the gas dissolves into the water: $CO_2 \Rightarrow CO_2^{aq}$. If the fraction of CO_2 in the gas were lowered there would be a reversal of the reaction and CO_2 would separate out of the aqueous phase. The exact ratio of products to reactants which leads to no net reaction is termed equilibrium and is defined by an equilibrium constant K, which is unique for a given reaction at a particular temperature and pressure.

$$K = \Pi[\text{Products}]_p^n / \Pi[\text{Reactants}]_r^n$$

where n_p is the number of products in a given reaction and n_r is the number of reactants in a given reaction. The Π's are the multiplicative equivalent of the summation sign, Σ. Brackets denote thermodynamic activities, which for salinities as high as seawater, can be converted to concentrations when the reactant or product is a dissolved species, using a technique such as the Debye-Huckel law which corrects for the salinity of the solution. (The thermodynamic properties of dissolved species are generally referenced to an ideal, salt-free solution). For simplicity, and because soil solutions and many groundwaters are generally non-saline, thermodynamic activities will be approximated below as equal to concentrations in mol/L (mol denotes moles = 6.022×10^{23} atoms; l denotes liter). The activities of pure solids and water are formally taken to be one.

Equilibrium reactions can be written between all of the likely chemical species in a soil or groundwater, and the likely chemical state of a contaminant (dissolved, chelated, sorbed, as a trace component of a sparingly soluble mineral, as a gas phase, etc. — see Figure 5.1) can thus be determined. Equilibrium constants for the reactions linking the various reactions are tabulated in a variety of thermodynamic databases (see e.g., Johnson et al., 1991; Morel and Hering, 1993; Stumm and Morgan, 1996). The reaction types themselves are briefly outlined below.

ION PAIRING

Ion pairs form between dissolved species because there is an electrostatic interaction between a positively charged cation and a negatively charged anion. For example, Cd^{2+} forms ion pairs with chloride ions:

$$Cd^{2+} + Cl^- \Leftrightarrow CdCl^+ \quad K_1 = [CdCl^+]/[Cd^{2+}][Cl^-] = 10^{1.98}$$

The more chloride and/or Cd that is dissolved in solution, the greater are the number of $CdCl^+$ ion pairs in solution. The actual number though is fixed by the relation above; i.e., by substitution, $[CdCl^+] = K_1[Cd^{2+}][Cl^-]$. If one were to somehow pull $CdCl^+$ ion pairs from solution and, for example, cause them to adhere to a mineral surface, the remaining Cd^{2+} and Cl^- would equilibrate, combining to re-establish the ratio fixed by the equilibrium constant for this reaction. Table 5.1 lists the common dissolved state (speciation) of elements in natural waters. However, in highly polluted environments having unusual solution compositions, significantly different species might be observed.

Two of the most important ion pairing reactions which prevail in soil and groundwaters involve protons (H^+):

$$H^+ + HCO_3^- \Leftrightarrow H_2CO_3 \quad K_2 = [H_2CO_3]/[H^+][HCO_3^-] = 10^{6.3}$$

$$H_2O \Leftrightarrow H^+ + OH^- \quad K_3 = [H^+][OH^-] = 10^{-14}$$

The first is the disproportionation of carbonic acid, H_2CO_3. It occurs as a consequence of the combination of carbon dioxide, from the atmosphere or soil reactions, with

Chemical Attenuation

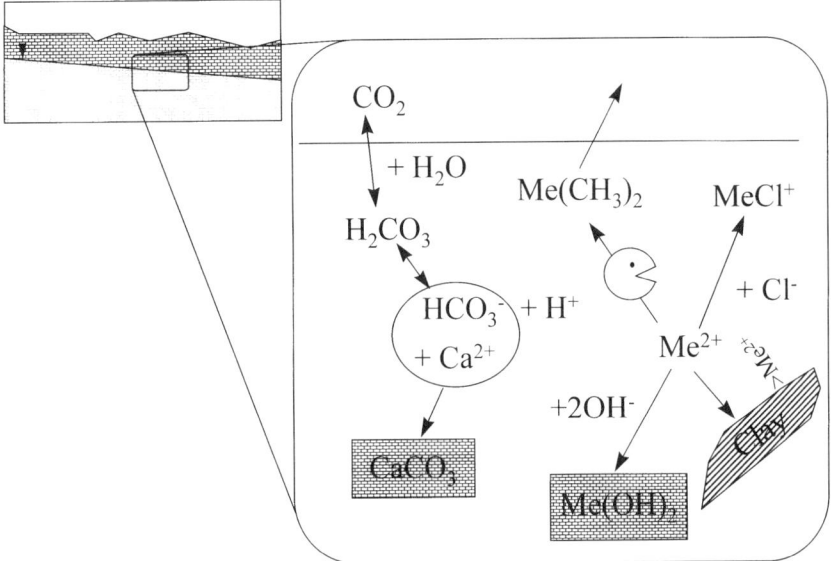

FIGURE 5.1 Reactions of metals in soils.

with water. The lower reaction is the disproportionation of water. The pH of a soil or groundwater is defined as pH = $-\log[H^+]$. Solutions with abundant protons are acidic and have low pH's. Low proton levels are typical of alkaline, high pH solutions. Most organic-rich soils have pH's in the range of 4 to 6. Drier soils containing calcium carbonate minerals have pH's of 7 to 8. Some alkali lakes have higher pH's than this. Soils where reduced sulfide minerals are weathering often have pH's less than 4.

GAS DISSOLUTION/EXSOLUTION

Vapor transport can cause a net transfer of contaminants from soils to the atmosphere where natural degradation processes (e.g., photolytic degradation, oxidation by OH radicals) are often more effective than those in soils. At the same time transport in the vapor phase can allow a contaminant to leap-frog ahead of a water-dissolved or separated liquid-phase plume. The partitioning of dissolved substances into soil vapor is described by Henry's law (Stumm and Morgan, 1996): $H_i = P_i/C_i$; where H_i is the Henry's law constant, which has units of atmosphere per mol/m³; P_i is the partial pressure (atmospheres) of the compound i over the fluid. C_i is the concentration of the compound i in the fluid (mol/m³). Henry's law simply says that higher concentrations of a compound in the fluid will lead to higher levels in the gas phase above the liquid as well. Note that H_i is substance specific. It is low for volatile molecules such as benzene (H_i = 0.0054 atmospheres/mol/meter³) and low for a nonvolatile pesticide like Lindane (H_i = 0.00000048 atmospheres/mol/meter³).

O_2 and CO_2 are two of the primary soil gases which dissolve into, and out of uncontaminated soil and groundwaters:

TABLE 5.1
Common Inorganic Species in Natural Waters

Condition	Element(valence)	Species
Hydrolyzed, anionic	B(III)	H_3BO_3, $B(OH)_4^-$
	V(V)	HVO_4^{2-}, $H_2VO_4^-$
	Cr(VI)	CrO_4^{2-}
	As(V)	$HAsO_4^{2-}$
	As(III)	$HAsO_2$
	Se(VI)	SeO_4^{2-}
	Mo(VI)	MoO_4^{2-}
	Si(IV)	H_4SiO_4
Free ions	Na(I)	Na^+
	Mg(II)	Mg^{2+}
	K(I)	K^+
	Ca(II)	Ca^{2+}
	Sr(II)	Sr^{2+}
	Cs(I)	Cs^+
	Ba(II)	Ba^{2+}
Complexation with OH^-, CO_3^{2-} HCO_3^-, or Cl^-	Be(II)	$Be(OH)^+$
	Al(III)	$Al(OH)_{3,s}$, $Al(OH)^{2+}$, $Al(OH)_4^-$
	Ti(IV)	$TiO_{2,s}$, $Ti(OH)_4$
	Mn(IV)	$MnO_{2,s}$
	Fe(III)	$Fe(OH)_{3,s}$, $Fe(OH)^{2+}$, $Fe(OH)_4^-$
	Co(II)	Co^{2+}, $CoCO_3$
	Ni(II)	Ni^{2+}, $NiCO_3$
	Cu(II)	$CuCO_3$, $Cu(OH)_2$
	Zn(II)	Zn^{2+}, $ZnCO_3$
	Ag(I)	Ag^+, $AgCl$
	Cd(II)	Cd^{2+}, $CdCO_3$, $CdCl^+$, $CdCl_2$
	La(III) — all lanthanides	$LaCO_3^+$, $La(CO_3)^{2-}$
	Hg(II)	$Hg(OH)_2$, $HgCl_4^{2-}$
	Pb(II)	$PbCO_3$, $PbCl^+$, $PbCO_{3,s}$
	Th(IV)	$Th(OH)_4$
	U(VI)	$UO_2(CO_3)_2^{2-}$, $UO_2(CO_3)_3^{4-}$

Note: s Denotes Solids.

After W. Stumm and J. Morgan, *Aquatic Chemistry,* John Wiley & Sons, New York, 1996. With permission..

$$O_2 \Leftrightarrow O_2^{aq} \quad K_4 = [O_2^{aq}]/P_{O2} = 10^{-2.9}$$

$$CO_2 \Leftrightarrow CO_2^{aq} \quad K_5 = [CO_2^{aq}]/P_{CO2} = 10^{-1.5}$$

aq is shorthand for aqueous and denotes a component dissolved in the aqueous, liquid phase.

Raoult's law (see Schwarzenbach et al., 1993) is generally used to describe volatilization of an organic from a separated organic phase, as opposed to a dissolved one. Raoult's law is $P_i = x_i P_{i,o}$; where P_i is the vapor pressure of the organic compound above the mixture. x_i is the mole fraction of the organic compound in the mixture, and $P_{i,o}$ is the partial pressure over the pure substance. Again, the latter varies greatly from compound to compound. Many of the halogenated alkanes and alkenes have relatively high vapor pressures. Trichloroethene has a $P_{i,o}$ of 0.0789 atmospheres. Benzene has a $P_{i,o}$ of 0.1 atmospheres. On the other end of the spectrum is DDT with a $P_{i,o}$ almost a billion times smaller at 1.3×10^{-10} atmospheres.

MINERAL GROWTH

A surprisingly limited number of minerals form from most soil solutions. They include $CaCO_3$, metal hydroxides — often of Fe(III), Al, Si, and Mn, — and sheet-like clay minerals. These minerals form relatively fast when the adjacent soil solution exceeds its carrying capacity for the respective mineral components. Again, the carrying capacity is fixed by thermodynamics, and depends on what else is in the fluid. Kinetics determines the rate at which a mineral appears, or, equivalently, the rate at which the dissolved mineral component disappears from solution. The carrying capacity of a solution for metals goes up when ion pairs are formed. For example, the carrying capacity of most soil solutions for Pb^{2+} is very low (<1 ppm, depending on the pH, alkalinity, and ionic strength) which is good for those who drink water from Pb-soldered pipes, or from aquifers recharged through soils contaminated with lead. EDTA, a common organic chelate, forms a relatively strong ion pair with Pb^{2+}, and, when present, causes the carrying capacity of a solution for Pb^{2+} to go up to match the amount of the EDTA-Pb^{2+} ion pair formed in solution. Mobilizing (or "chelating") ligands (chemical partners) are important because they usually increase the metal-carrying capacity of soils and groundwaters. Some important ones are organic acids, formed by natural soil processes, methyl ions formed by bacteria, associated contaminant solvents such as EDTA, and, for many radio-nuclides, CO_3^{2-}. Just as mobilizing ligands generally make it unlikely that a given metal will drop from solution into a freshly formed mineral, metals associated with mobilizing ligands often have lower net charge, so they are less likely to be sorbed.

Many metals having a +2 charge are least soluble at above neutral pH (7) because they form carbonate mineral phases and/or become trace constituents in $CaCO_3$. This is true for Pb, Cd, and, to an extent, Cu. Radioactive americium also forms carbonate minerals at alkaline pH which limits its aqueous transport. Hg does not form a carbonate phase for the most part. Because most natural reducing environments contain hydrogen sulfide, it follows that a host of divalent metals will be immobilized as metal sulfide minerals in oxygen-poor waters. CrO_4^{2-} is reduced to Cr^{3+} which forms a relatively insoluble hydroxide. U and Np are reduced and insoluble under such conditions as well. (Anionic contaminants such as CrO_4^{2-} and I^- are less likely to form insoluble solids in most cases).

The identity of the solid which forms is important as it can control the concentration of the given metal in groundwater, which in turn largely determines the bioavailability of the metal. At the same time different metal-containing solids dissolve at different rates when ingested, making available their toxicity at greater or lesser rates. Identifying the primary metal-containing solid is consequently quite important.

The extent to which nonionic and nonpolar (hydrophobic) organic compounds partition from a separated phase or organic solid into soil solutions, or from the latter into the soil atmosphere can be roughly predicted (see Chapter 6). Partitioning of organic molecules from the separated phase or organic solid into solution is often expressed in terms of octanol-water partitioning coefficients, K_{ow}. The latter measures the tendency of molecules to separate from water into octanol, an organic solvent. Increasing salt content of the solution typically decreases the solubility of the organic. Temperature changes can increase or decrease organic solubilities depending on the molecule and the temperature range. Organics with ionizable functional groups are generally more soluble, depending on their degree of ionization. The thermodynamics and molecular origins of organic solubilities are outlined in Schwarzenbach et al. (1993).

One of the more relevant mineral growth reactions is that of iron hydroxide.

$$Fe^{3+} + 3OH^- \Leftrightarrow Fe(OH)_3 \quad K_6 = 1/[Fe^{3+}][OH^-]^3 = 10^{38}$$

$Fe(OH)_3$ and its less hydrated equivalents, are important electron acceptors for the degradation of organic contaminants under anaerobic conditions. Iron (hydr)oxide minerals are also some of the most reactive sorbers of heavy metals in soils. Note that, by the reaction above, at a soil pH of 7 the total amount of Fe^{3+} left in solution $[Fe^{3+}] = [OH^-]^3 10^{38} = 10^{-17}$ mol/l, which is incredibly small. In fact the amount of iron that is measured in a solution in equilibrium with $Fe(OH)_3$ is several orders of magnitude higher due to the presence of a number of ion pairs which Fe^{3+} forms with hydroxyls (OH^-):

$$Fe^{3+} + OH^- \Leftrightarrow Fe(OH)^{2+} \quad K_7 = [Fe(OH)^{2+}]/[Fe^{3+}][OH^-] = 10^{11.8}$$

$$Fe^{3+} + 2OH^- \Leftrightarrow Fe(OH)_2^+ \quad K_8 = [Fe(OH)_2^+]/[Fe^{3+}][OH^-]^2 = 10^{21.3}$$

$$Fe^{3+} + 3OH^- \Leftrightarrow Fe(OH)_3^{aq} \quad K_9 = [Fe(OH)_3]/[Fe^{3+}][OH^-]^3 = 10^{28.4}$$

$$Fe^{3+} + 4OH^- \Leftrightarrow Fe(OH)_4^- \quad K_{10} = [Fe(OH)_4^-]/[Fe^{3+}][OH^-]^4 = 10^{33.0}$$

By manipulating these equations we can calculate the pH-dependence of the distribution of Fe-OH species, and solubility of the solid $Fe(OH)_3$. (The $Fe(OH)_3^{aq}$ in Reaction 9 is a dissolved ion pair, not a solid, though the chemical formula is the same). The technique for doing this is described in a number of aquatic chemistry textbooks (Stumm and Morgan, 1996). In general, geochemists now rely on computer programs to simultaneously solve the set of equations.

The solubility diagram for $Fe(OH)_3$ is shown in Figure 5.2. At low pH, there are relatively few hydroxyls in solution and, consequently, few Fe-hydroxyl ion pairs. At high pH, where hydroxyls are abundant, levels of $Fe(OH)_4^-$ are high. At 6 <pH

Chemical Attenuation

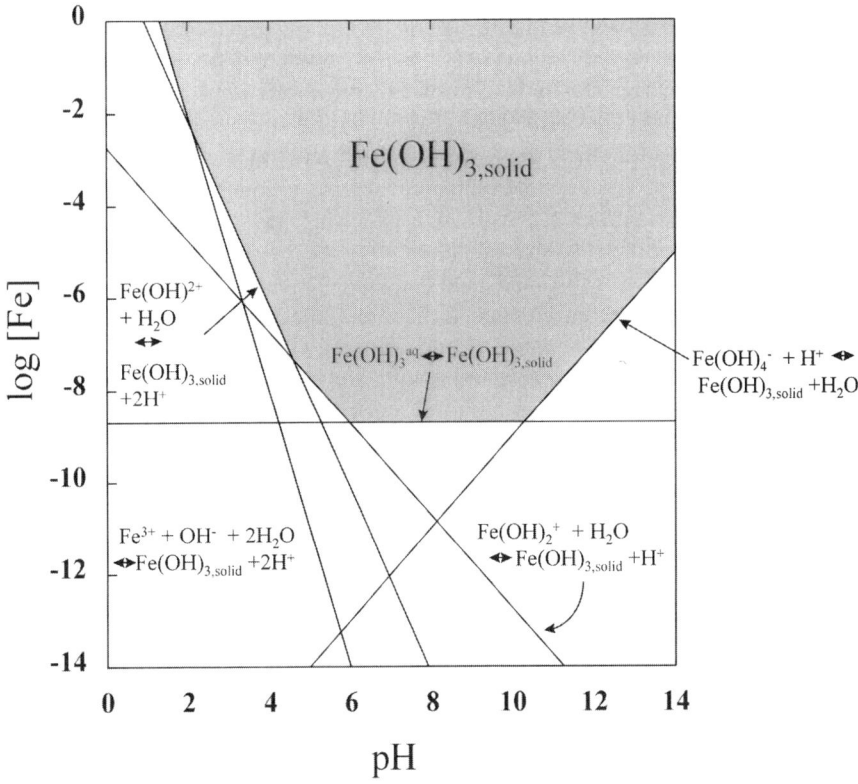

FIGURE 5.2 Solubility diagram for Fe(OH)$_3$.

<10 most of the iron in solution is in the form of Fe(OH)$_3^{aq}$, which has no net charge. As such it is not likely to be attracted to any other charged aqueous species. At the same time it finds itself out of place among the water molecules, which themselves are polar in that they have an assymetrical charge distribution. Consequently, iron at this pH tends to unmix from the water to form its own solid. The solubility minimum for Fe^{3+} is centered at pH ~8.5 and is the region where neutral Fe-OH species form. The same pH behavior defines the solubility for Al^{3+}, though in the latter case the solubility minimum is at pH ~6.5. Higher pH's (pH 9 to 11) are typical of the solubility minimum for Pb, Cd, Zn, Ni, as well as iron in its more reduced, divalent state. Solutions whose compositions plot along the line separating the stability field for Fe(OH)$_3$, the shaded portion (and containing no other complexing agents), from the rest of the graph are at equilibrium with respect to iron in solution and Fe(OH)$_3$. Such solutions are termed "saturated". If they contain more iron than the graph would predict at a given pH they are termed "supersaturated" or "oversaturated". If they contain less iron they are termed "undersaturated" with respect to Fe(OH)$_3$.

If one were to add Fe^{3+} to a Fe(OH)$_3$-saturated solution buffered to pH 8.5, it would all become Fe(OH)$_3$ — none would be added to solution. If, however, the solution pH were lowered to say 4, Fe^{3+} would dissolve from the $Fe(OH)_3$ solid to maintain the border shown in Figure 5.2. At pH 4, Fe^{3+} levels must be around $10^{-6.5}$ mol/L to remain at equilibrium. At pH 8.5, Fe^{3+} need only be ~$10^{-8.7}$ mol/L. The increase in solubility caused by the downward shift in pH would be made up for by dissolution of some of the $Fe(OH)_3$. In reality there are further complications to the Fe-OH system. As $Fe(OH)_3$ ages it often dehydrates to from a variety of iron (hydr)oxides such as goethite and lepidocrocite, and ultimately hematite. Trivalent iron minerals are ubiquitous in oxygen-rich soils where they provide a characteristic reddish-orange tinge.

In general, the iron is insoluble, and consequently immobile, unless it becomes reduced (that is acquires an electron from some other compound) to the divalent state. Often the other compound is an organic compound and the whole transaction is directed under oxygen-free conditions by a soil microorganism. When trivalent iron is changed to divalent iron it becomes much more soluble. Recall that divalent cations have solubility minima at much higher pH than most pH's typical of groundwater. Consequently, hydroxides made of Fe^{2+} are quite soluble at neutral and lower pH's. Divalent iron is therefore a useful fingerprint of oxygen-free, anaerobic conditions. The transformation of trivalent iron in a solid to divalent iron in solution is termed a reductive dissolution reaction. The dissolution part is pretty straightforward. It is the opposite of mineral growth. Where there are fewer components of a mineral in solution than thermodynamics says there should be, the tendency is for the mineral to dissolve to re-achieve equilibrium. A shift in pH from 8.5 to 4 would undersaturate the solution in contact with $Fe(OH)_3$ and cause dissolution of the latter. If there was less $Fe(OH)_3$ to begin with than was needed to re-achieve saturation (equilibrium) all of the $Fe(OH)_3$ would dissolve and the amount of iron in solution would be less than that predicted by thermodynamic equilibrium between the solution and the solid. Instead iron levels would reflect the balance between supply and demand (dilution + sorption).

If trivalent iron at the $Fe(OH)_3$ surface, or in solution, were reduced to divalent iron the effect would be to momentarily undersaturate the adjacent solution. This in time would cause further $Fe(OH)_3$ dissolution. Analogous electron transfer reactions allow (hydr)oxides of Mn^{4+} to be used for the anaerobic oxidation of organic compounds.

ELECTRON TRANSFER REACTIONS

Electron transfer reactions are fundamentally important and include such everyday phenomena as as the rusting of iron metal, wherein Fe° donates electrons to atmospheric oxygen to form red-colored, iron oxide and hydroxide coatings. Electron transfer reactions are critical to many of the transformation steps of soil and groundwater contaminants and, consequently, deserve further explanation. When soluble divalent iron is turned to insoluble, trivalent iron, an electron is passed onto some other compound by the Fe^{3+}:

$$Fe^{2+} \Leftrightarrow Fe^{3+} + e^-$$

Chemical Attenuation

e^- is the electron in the reaction above and the Fe^{3+} usually ends up on a mineral surface once formed, as opposed to being in solution. Although we could write the equilibrium constant expression for the reaction above it is probably better not to, as it would require a term describing the thermodynamic activity of the electron, something which has a very short half-life in aqueous solutions. Instead reactions are written so that the e^- doesn't formally appear. This is done by matching each electron donor reaction, like that above, with an electron acceptor reaction. If the electron given up by Fe^{2+} was added to atmospheric oxygen, the sum of the electron donor and electron acceptor reactions would be summed as follows:

$$4Fe^{2+} \Leftrightarrow 4Fe^{3+} + 4e^-$$

$$4H^+ + O_2 + 4e^- \Leftrightarrow 2H_2O$$

$$4H^+ + 4Fe^{2+} + O_2 \Leftrightarrow 2H_2O + 4Fe^{3+} \quad K_{11} = [Fe^{3+}]^4/[Fe^{2+}]^4 P_{O2} = 10^{31.1}$$

The important point here is that in the overall reaction a compound accepts electrons and is reduced in charge, while another compound loses electrons and is oxidized to a higher charge species. Electrons are the currency of oxidation-reactions, though they never appear in the transactions. A crude but useful analogy of the situation is that of an economy in which every good or service posesses a dollar value but all transactions are done by barter with no currency changing hands. The dollar value would provide a critical point of reference for each exchange. The reference point in electron transfer reactions is the system E_H.

Figure 5.3 shows schematically where the various electron transfer (redox) reactions occur as a function of system E_H. E_H has units of volts and is equal to $-2.303RT\log[e^-]/F$; where R is the gas constant, T is temperature (in Kelvins), and F is Faraday's constant (23.06 kcal per volt gram equivalent). Combining all these terms; at 25°C E_H is equal to $-0.059\log[e^-]$. E_H measures the activity of electrons in a soil or groundwater. When the E_H is low, conditions favor high electron activity and redox-sensitive compounds tend to be in their most reduced (electron-rich, low charge) state: e.g., Fe^{2+} would be more common than Fe^{3+}, and oxygen would be encountered as O^{2-}, rather than uncharged O_2. When the E_H is high, redox-sensitive compounds tend to be in their highest valence states. Obviously, the E_H at which electrons are transferred differs from compound to compound. In water-logged soils the E_H can approach zero volts. In well-mixed surface waters it is between 0.5 and 1.0 volts. Knowing the E_H (redox-state) of aquifers and soils is critical to assessing the fate of most redox-sensitive compounds. Some, like iron, are very immobile in their oxidized state. Others, such as Cr, are most insoluble and immobile under reducing conditions. Redox conditions also often dictate the extent and the rate at which organic compounds are broken down. For example, PCE can be converted sequentially to TCE, DCE, and vinyl chloride under reducing conditions. PCE and TCE are common industrial solvents — DCE is not. Whenever the latter shows up in groundwater, it may be a breakdown product of the former. Vinyl chloride is quite toxic and generally resists rapid breakdown under most reducing conditions. Instead, more oxidizing conditions are required for the ultimate conversion of vinyl chloride

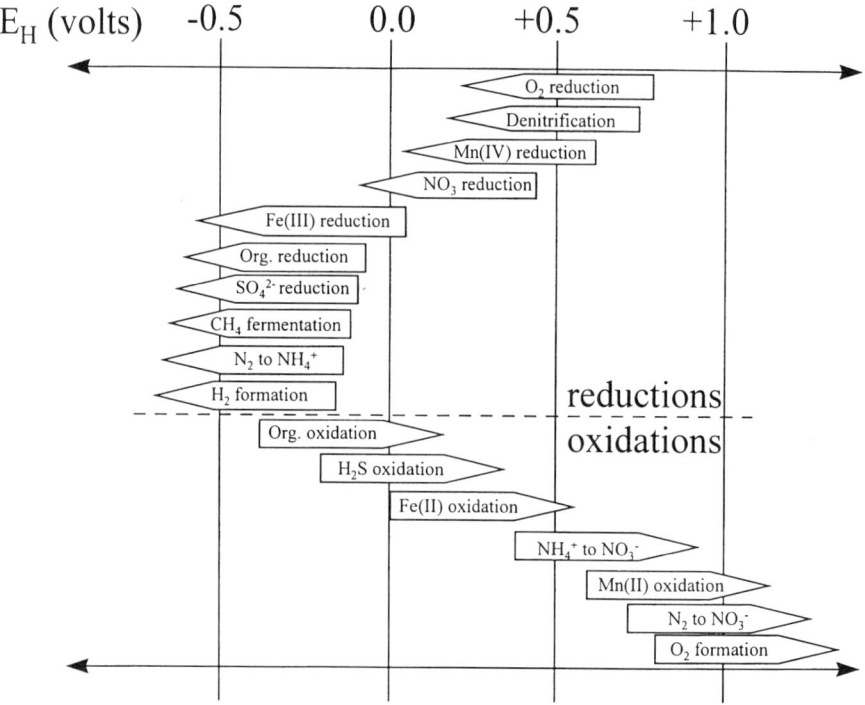

FIGURE 5.3 Electron transfer reactions as a function of system E_H. (After W. Stumm and J. Morgan, *Aquatic Chemistry*, John Wiley & Sons, New York, 1996. With permission.)

to carbon dioxide and chloride. Ethene is a breakdown product of TCE and PCE degradation under reducing conditions that is otherwise an uncommon constituent of groundwaters.

In theory one should be able to measure E_H in the field with a platinum electrode. In fact what the latter tends to measure is the E_H of only those redox couples which are capable of reacting with the electrode, and many important redox reactions don't register at all with field electrodes. Consequently, system redox state must be indirectly estimated by the presence of the various redox-sensitive compounds (see discussion of Stumm and Morgan, 1996). Chappelle and co-workers have shown that H_2 is an effective monitor of system redox conditions for aerobic processes (e.g., Chapelle, 1993). H_2 is an important intermediate product during anaerobic microbial metabolism, and its concentration is characteristic of the different electron-accepting processes (see Table 5.2).

ADSORPTION

Adsorption is simply the "sticking" of a molecule to a mineral surface. The broken chemical bonds which make up a mineral surface take up dissolved molecules from solution to replace their missing neighbors. Mineral surfaces also acquire contaminants from solution by providing a refuge for non-polar organic molecules wishing

TABLE 5.2
H_2 Levels and Redox Conditions

Redox regime	H_2 level
Nitrate reduction	<1 nM
Fe(III) reduction	0.2 to 0.8 nM
SO_4^{2-} reduction	1 to 4 nM
Methanogenesis	5 to 15 nM

Note: nM = nanomol/l

After Chapelle, 1993.

to unmix from water, a polar solvent. While the surface sometimes gains stability in the process, the once-dissolved molecule loses its mobility in the bargain. Adsorbed molecules often find their way into "holes" at mineral surfaces, from which it is often difficult for them to be extracted (see below). The affinity of a given contaminant for the ambient soil mineralogy goes a long way toward determining whether the contaminant moves at velocities approaching that of the soil fluid (no sorption), or not at all (strong sorption). Adsorptive affinity depends on both the contaminant and soil mineralogy, and both vary widely in their contribution to sorption.

There is a continuum between sorption and mineral precipitation. As a sorbate covers a mineral surface to the thickness of one molecular layer, and then more, the sorbed molecules begin to associate and polymerize, sometimes forming new minerals in the process. In other words, at high surface loading there is a blending between sorption and mineral growth. If the growing mineral is made up of the contaminant of interest, or even if the contaminant is only a trace constituent in the mineral, limits are set on the amount of contaminant remaining in solution, which might subsequently be transported by fluid movement. The actual amount depends on the solubility of the mineral.

The three most reactive sorbers in most near-surface soils are soil organic matter, iron (hydr)oxides, and clays. The first of these is extremely heterogeneous and can be leaf litter, decaying roots, microbiota, and the like. Decay of vegetation leads to the formation of humic substances. The latter is defined as the fraction that can be extracted from a soil using a high-pH leach solution. Once the leachate is acidified, "humic" acids precipitate out, leaving "fulvic" acids in solution. In soils both components can sorb metals, though fulvic acids, due to their soluble nature are more likely transporting agents. The organic matter which can't be extracted in the first place is referred to as "humin". Note that non-polar components of each organic fraction might sorb non-polar organic contaminants from solution.

Iron hydroxides have very high surface areas and are seen to be very powerful agents for sequestering metals. Below about pH 8, $Fe(OH)_3$ some surface sites are positively charged and capable of sorbing anions. Above about pH 6, $Fe(OH)_3$ can scavenge cations, and particularly so above pH 9. At low pH, the high positive

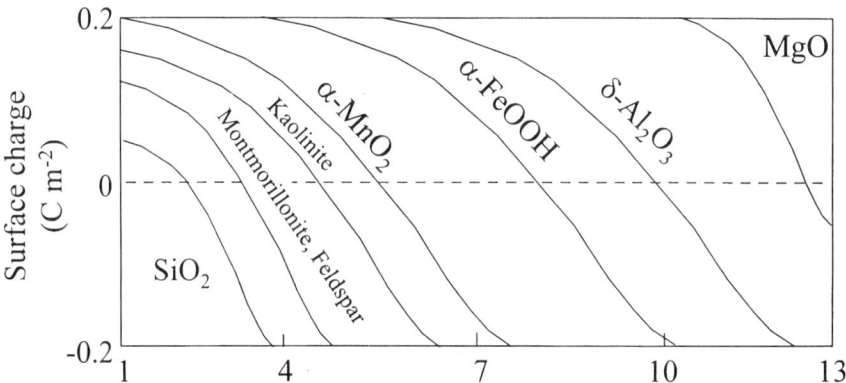

FIGURE 5.4 pH-dependent surface charge of common soil minerals. (After W. Stumm and J. Morgan, *Aquatic Chemistry,* John Wiley & Sons, New York, 1996. With permission.)

surface charge repels cations. Likewise, at very high pH (pH >10) the highly negative surface charge tends to repel anions. The pH-dependent surface charge of minerals varies widely and depends on the affinity of the broken metal-oxygen bonds exposed at mineral surfaces for protons and hydroxyls. Figure 5.4 shows the general distribution of pH-dependent surface charge for a number of common soil minerals.

Clay minerals are sheet-like minerals which sorb metals in two ways (see Figure 5.5). Some metals are loosely bound at the outer edges of the plates. Some are bound to negatively charged sites between the plates that come about from the substitution of, for example, trivalent aluminum for quadrivalent silicon in the mineral lattice. The heterovalent substitution consequently imparts a "permanent" negative charge to basal planes. Consequently, these sites do not show a reversal in polarity in normal groundwater pH solutions. Movement of a metal ion to a ledge opening at a surface is also shown in Figure 5.5. Extensive examinations of mineral surface chemistry and its controls on transport are found in (Brady and Zachara, 1996; Davis and Kent, 1990; Dzombak and Morel, 1990; Stumm and Morgan, 1996).

MINERAL DISSOLUTION AND CONTAMINANT AVAILABILITY

If a metal contaminant exists as a solid in soils, its availability to groundwaters depends on both its solubility, which determines whether or not it will dissolve into soil or groundwaters, and, assuming that the solution is undersaturated, its rate of dissolution. Thermodynamics indicates when a mineral is unsaturated, and, therefore, when there is an energetic driving force for dissolution. Sometimes, as in the case of iron hydroxide, dissolution rates are very rapid, and the mineral dissolves rapidly enough that solutions stay very close to saturation (equilibrium). For many other materials of environmental importance (slags, radioactive glasses, and sulfide minerals found near smelters) dissolution rates are much slower. The thermodynamics of the process don't provide much insight about contaminant release. For these

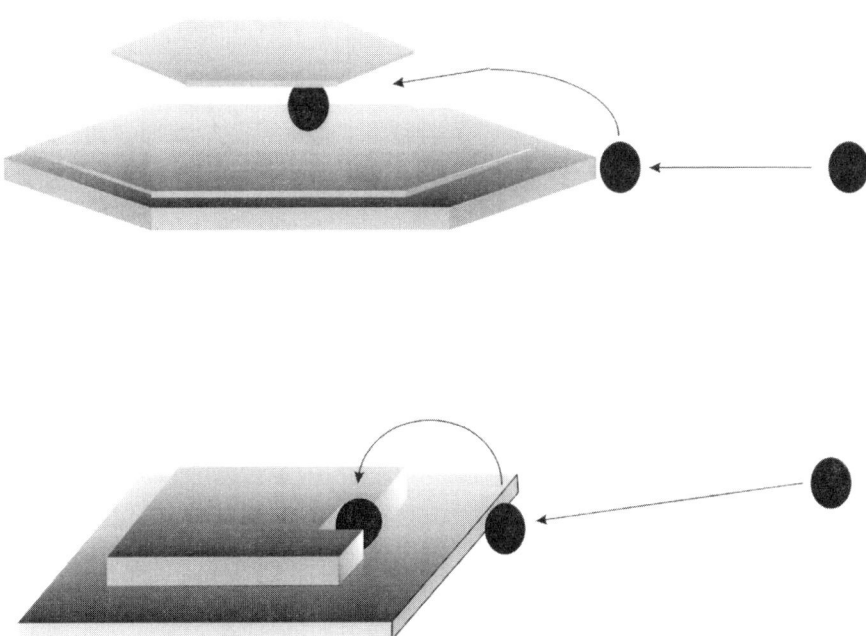

FIGURE 5.5 Migration of sorbed molecules to strong-binding sites.

slow-dissolving materials, it is the dissolution rate which controls contaminant release.

This being said, it is not always easy to predict dissolution rates of minerals in soils (see Brady and Zachara, 1996). Generally, the more undersaturated a solution is with respect to a mineral, the greater will be its rate of dissolution. Rates of mineral dissolution are cast in terms of moles of a mineral dissolved per cm^2 of mineral surface area exposed per second ($mol/cm^2 s$). Silicates (materials made up largely of SiO_2 groups) dissolve faster than carbonate minerals (those made up of metals + CO_3^{2-} groups). Metal sulfide dissolution rates depend largely on the availability of electron acceptors (O_2, Fe^{3+}, etc.) (Moses et al., 1987). Dissolution rates of many phosphate minerals are hard to measure, in part because they are so insoluble that undersaturation is not always easy to achieve.

Figure 5.6 shows dissolution rates of a number of silicates as a function of pH. Note that between pH 5 and 8 dissolution rates for any given mineral are insensitive to pH. Above pH 8 dissolution rates roughly double for every unit increase in pH, presumably reflecting hydroxyl-catalyzed dissolution. These rate-accelerating hydroxyls are thought to be present at the mineral surface (Brady and Walther, 1989). Below pH 5 rates increase exponentially with pH, though the pH-dependence of this increase varies from silicate to silicate. Presumably, the rate increase reflects the presence of rate-promoting protons at the various mineral surfaces (Stumm, 1995).

Adsorbed species other than protons and hydroxyls are observed to cause similar large-scale amplifications of dissolution rates. Organic acids are thought to accelerate the weathering of many silicates (Cochran and Berner, 1993; Graustein et al., 1977;

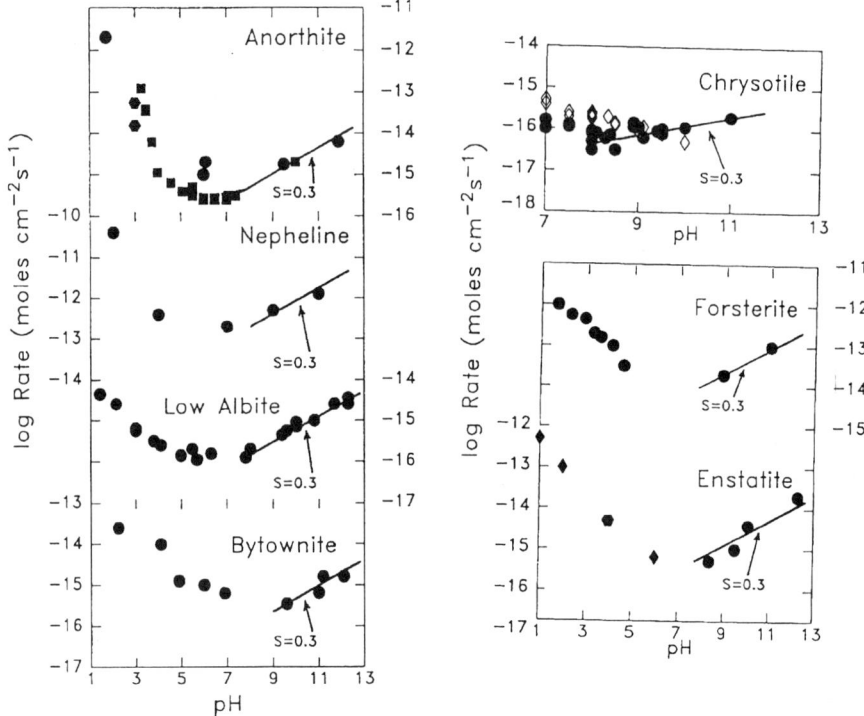

FIGURE 5.6 Silicate dissolution rates at 25°C as a function of pH. (From Brady, P.V. and Walther, J.V., *Geochim. Cosmochim. Acta,* 53, 2823, 1989. With permission from Elsevier Science Ltd., Kidlington, U.K.)

Welch and Ullman, 1993). High salt concentrations may also amplify the dissolution rates of some silicates at high pH (Dove, 1994). Figure 5.7 shows dissolution rates of the more common carbonate minerals as a function of pH at 25°C. At high pH, carbonate minerals in most soils are relatively insoluble. However, note that, like the silicates, the absolute values of the rates vary widely from mineral to mineral. This is due to inherent differences in the strengths of the various metal-oxygen bonds at the respective mineral surfaces (Brady and House, 1996).

Just as some ligands (e.g., organic acids), once sorbed to mineral surfaces, are capable of amplifying mineral dissolution rates, some ligands also retard, or slow down, corrosion rates once sorbed. The most effective of these is probably phosphate, which sorbs strongly to, and retards the dissolution of a wide variety of soil minerals. Large organic macromolecules are capable of doing the same thing.

GEOCHEMICAL MODELING

Thermodynamics allows one to predict what reactions should occur in soil solutions, and there are a number of computer codes, such as EQ3/6 (Wolery, 1983), REACT (Bethke, 1994), WATEQ (Plummer et al., 1976), and MINTEQ, which do this, given

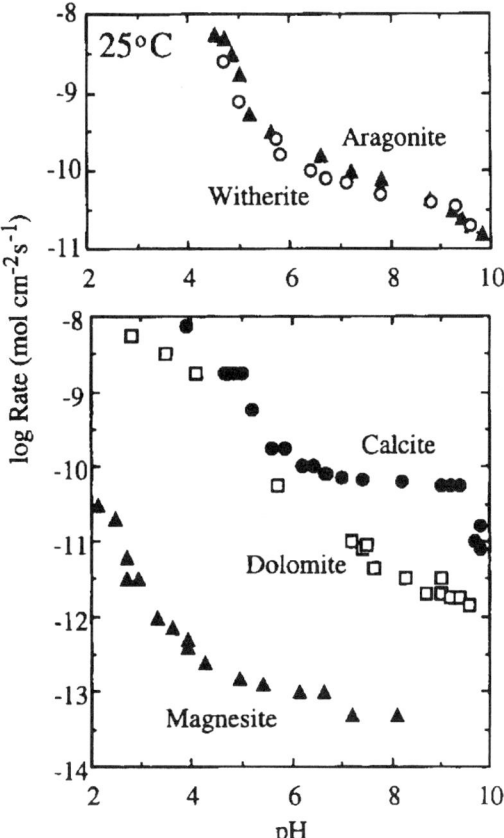

FIGURE 5.7 Metal carbonate mineral dissolution rates. (After Chou, L. et al., *Chem. Geol.,* 78, 269, 1989. From Brady and House, 1996. With permission.)

a knowledge of the chemical makeup of the groundwater. All are based on the same chemical and physical principles of mass balance and mass action (chemical equilibrium). Nevertheless, the results the various code predictions often differ due to the fact that thermodynamic databases are not identical, and different corrections are used to account for non-ideality of solutions. Furthermore, predicted mineral assemblages and fluid speciation are often at odds with what is observed in the field for a number of additional reasons. Some codes don't take account of sorption and/or ion exchange, while others can't model the kinetics of geochemical reactions. In general, ion pairing reactions are fast and go to equilibrium in seconds or less. The transfer of gases across the fluid-soil interfaces is generally somewhat slower. Dissolution and growth of many minerals is sufficiently sluggish that geochemical codes which assume equilibrium between minerals and fluids (and gases) often predict outcomes quite different from reality.

Some "tinkering" is almost always required to make the models and reality consonant. This generally involves suppressing the appearance of minerals which thermodynamics favors but kinetics prevents. Only recently have codes been modified to model the growth of some common soil minerals such as clays, which are kinetically favored. Unfortunately, many electron transfer reactions are kinetically slow. Equilibrium thermodynamics provide only semi-quantitative clues to the fates of redox-sensitive contaminants, the most important being xenobiotic (synthetic) organics.

Clearly, the water that moves between soil minerals and in pores at depth is delicately balanced in a chemical sense. In all but the saltiest of brines, soil and groundwaters contain far less dissolved material than the minerals they are intermingled with. A wealth of evidence indicates that the chemical composition of most groundwaters is, in effect, defined by reactions with the aquifer solids (Hem, 1989). When groundwaters react with the adjacent rock or soil, only rarely does a single chemical change its concentration. Instead, a whole series of interlocking reactions tend to ensue. A useful example of this interdependence is the aerobic biodegradation of benzene (see Table 6.1 in Chapter 6). While oxygen is consumed, CO_2 is produced (along with water). CO_2 will then combine with water to form H_2CO_3, carbonic acid, which causes a drop in pH and a rise in bicarbonate levels upon disproportionation of carbonic acid: $H_2CO_3 \Leftrightarrow H^+ + HCO_3^-$. The added acidity may subsequently be consumed by reaction with calcium carbonate minerals: $CaCO_3 + H^+ \Leftrightarrow Ca^{2+} + HCO_3^-$, and so on. For a plume of biodegrading BTEX we should see levels of the latter drop. But we should also see changes in bicarbonate, Ca (if calcite is present in the rocks or soils), and pH relative to BTEX-free sites, as well as any trace elements (e.g., Mg^{2+}, Sr^{2+}, Ba^{2+}, etc.) that might have been associated with the limestone. At the same time, fractionation of the various isotopes of carbon, oxygen, and hydrogen are also potential indicators of degradation mechanisms, and rates.

ISOTOPIC MONITORING OF CONTAMINANT DEGRADATION

Isotopes of carbon have received the most attention as potential tools for fingerprinting natural attenuation of organic contaminants. Three isotopes of carbon exist in nature: ^{12}C, ^{13}C, and ^{14}C. The last of these is present in relatively small quantities and is radioactive, having a half-life of 5730 years and decaying ultimately to ^{14}N. ^{14}C is produced in the upper atmosphere by interaction of cosmic rays with ^{14}N (as well as by above-ground nuclear weapons testing). ^{12}C and ^{13}C are stable isotopes and make up, respectively, 98.89 and 1.11% of non-radiogenic carbon. The ratio of ^{13}C to ^{12}C in a group of carbon atoms is changed by chemical reaction. It is this change (fractionation) which can be used to provide information about chemical reactions occurring in the subsurface. Changes in isotopic ratios are calculated relative to the $^{13}C/^{12}C$ ratio in an internationally accepted standard, the PeeDee Belemnite (PDB). Specifically, the difference between a sample and the standard is calculated in per mil (‰) as:

$$\delta^{13}C = 1000 \times \{[^{13}C/^{12}C]_{sample} - [^{13}C/^{12}C]_{PDB}\}/[^{13}C/^{12}C]_{PDB}$$

$\delta^{13}C$ is positive when a sample has a higher $^{13}C/^{12}C$ ratio than the PeeDee Belemnite and negative when the ratio is lower. $\delta^{13}C$ of bicarbonate in the ocean is around 0‰. $\delta^{13}C$ of atmospheric carbon dioxide is around 7 to 10‰. When the latter is abstracted from the atmosphere and combined with water during photosynthesis to form carbohydrate material there is a fractionation downward of around 20 to 25‰, and terrestrial vegetation consequently has a $\delta^{13}C$ of about –25‰. Soil CO_2 comes from biologic degradation of this material and its $\delta^{13}C$ is around –20‰. The carbon in carbonate minerals is heavier (having more ^{13}C), generally having a $\delta^{13}C$ of around 0‰. Non-methane petroleum hydrocarbons have a $\delta^{13}C$ of approximately –30‰. The $\delta^{13}C$ of methane varies widely between –110 to –13‰ (Fuex, 1977).

^{13}C and ^{12}C, in part because of the differences in their masses, don't react identically. Consequently, when a mixture of the two reacts to form a new compound, a difference in the isotopic ratio is observed between the reactant and the product. This fractionation, if known beforehand, can be used in combination with the $\delta^{13}C$ signature of reaction products to gain information about what form the carbon was in to begin with.

One of the most succesful applications of carbon isotopes to natural attenuation processes has been their use to quantify *in situ* biodegradation of fuel hydrocarbons (Aggarwal and Hinchee, 1991). The approach relies on the measurement of carbon isotopes in the breakdown products, namely soil CO_2. Although measurements of CO_2 levels have often been used to implicate biodegradation, the fact that there are many sources of soil CO_2 (root respiration, breakdown of natural organic matter, calcite formation, diffusion in from the atmosphere, etc.) makes it difficult to determine what fraction actually points to biodegradation in the subsurface. Aggarwal and Hinchee (1991) examined the aerobic degradation of fuel hydrocarbons at two U.S. Air Force Bases (Hill, Tyndall) and one U.S. Naval Air Station (Patuxent River), noting first of all that the ^{13}C of soil CO_2 collected above the aerobically degraded, naturally occurring organic matter was close to that of the organic matter itself, hence there appears to be no discernible fractionation during the biodegradation of naturally occurring organic matter. The naturally occurring background $\delta^{13}C$ depends on the vegetation and carbon fixation type. Soil CO_2 levels were highest in areas where degradation of fuel hydrocarbons was most extensive. For these samples the $\delta^{13}C$ in soil CO_2 was measured to be 3 to 5‰ lower than that of soil CO_2 from uncontaminated sites, and there was a clear correlation between CO_2 production by biodegradation and isotopic fractionation. Consequently measurement of isotopic fractionation is seen to be a useful tool for demonstrating aerobic degradation of fuel hydrocarbons, and for discriminating between potential degradation substrates (see example 1 of Van de Velde et al., 1995)). In addition to outlining a sampling protocol, Van de Velde et al. (1995) identify the primary benefits of stable carbon isotope analysis as:

1. Confirmation of biodegradation: stable carbon isotope analysis can provide evidence of partial or complete mineralization of the targeted substrate.
2. Reduced analytical costs: stable carbon isotope analysis is relatively inexpensive (30 to 65 dollars per sample) compared to other analyses.

3. Fewer analyses: Fewer samples are needed for a given site, primarily because heterogeneity is less of a problem for soil gas samples than for soils, sludge, and waters.
4. Reduction of sampling costs: Fewer samples and the simpler collection procedure for gas samples reduce the amount of time necessary to collect samples.

All this being said, there may be some real down sides to routine collection of isotopic data. A dissenting view.

> Stable carbon isotope analysis requires an additional level of site characterization that is more costly, subject to greater interpretation, and less uniform from site to site. For example, the ^{13}C signature of the soil would have to be characterized for each site. The heterogeneity of a soil gas sample is less than other media but site-specific geology controls the likelihood of collecting a soil gas sample that reflects the contaminant levels in the groundwater. There are many more problems associated with soil gas sampling than with groundwater sampling. An analysis of BTEX in groundwater costs $85 and is a direct measure of the medium of concern, and is preferred to a measurement of an indicator parameter (^{13}C) from a secondary medium (soil gas).

(Personal Communication, Dr. R. T. Hackler, Ecology and Environment, Inc., Chicago, Illinois, 1/24/97).

Revesz et al. (1995) showed at the USGS Bemidji site (see below) that measurement of $\delta^{13}C$ in soil methane can specifically identify the source of the latter (acetate fermentation vs. CO_2 reduction by H_2). Trust et al. (1995) measured $\delta^{13}C$ values of CO_2 above PAHs being degraded by *Sphingomonas paucimobilis* and observed that the $\delta^{13}C$ of the contaminant being degraded (fluoranthene) was unchanged over time. The $\delta^{13}C$ of the CO_2 (and, over time, bacterial nucleic acids) approximated that of the fluoranthene, suggesting that stable carbon isotope analyses might be useful in the degradation of PAHs as well. Isotopic monitoring of the breakdown of chlorinated solvents, such as PCE and TCE, is still in its infancy; but there is considerable impetus for its development. Not only do chlorinated solvents make up a large fraction of the contaminants in groundwaters, they also can end up in a variety of locations in soils and groundwaters. Volatilization, adsorption, dilution, and dispersion, as well as biodegradation, can cause decreases in their levels. The net result is that natural attenuation of chlorinated solvents through biodegradation might be masked by the other sinks. This, ultimately, makes it difficult to predict the controls on, and extent of, contaminant drawdown over time. As with fuel hydrocarbons, stable isotopes might potentially be used to quantify the fraction of chlorinated solvent drawdown specifically attributable to biodegradation.

For the breakdown of PCE, TCE, DCE, and vinyl chloride, the focus has been on quantifying fractionations in the isotopes of carbon and chloride, and under aerobic conditions, in molecular oxygen. Although from mass considerations alone one might expect large fractionations in the stable isotopes of hydrogen during biodegradation, a number of instrumental obstacles prevent their measurement (C. B. Douthitt, personal communication). There are two natural isotopes of chlorine:

^{35}Cl and ^{37}Cl, which make up, respectively, 75.53 and 24.47% of the total. Mathematically, the standardization is directly analogous to that listed above for carbon isotopes, though ratios are referenced relative to a standard defined by their ratio in seawater. Vanwarmerdam et al. (1995) made one of the first attempts to discern a pattern in chlorine and carbon isotope ratios in chlorinated solvents, specifically, in PCE, TCE, and 1,1,1-trichloroethane obtained from a variety of manufacturers. Generally, values of δ^{37}Cl of the samples were somewhat similar and ranged between –3.5 and +6.0‰; δ^{13}C ranged from –37.2 to –23.3‰. Recent work by Sturchio and co-workers (Holt and Sturchio, 1996) has focused on indentifying the effect of biodegradation on chlorine isotope ratios.

ADSORPTION AND TRANSPORT

Most transport codes used to model and predict the transport of contaminants in soils model adsorption by the "retardation factor" approach:

$$R_f = V/V_c \approx 1 + K_d \cdot (\rho/\theta)$$

where R_f is the retardation factor, V is the advective speed of the bulk fluid, V_c is the travel speed of the contaminant front or peak, K_d is the solid/fluid distribution coefficient in mL/g (see below), ρ is the bulk soil or rock density, and θ is again the porosity. The fluid is assumed to have a density of 1 g/mL. If $R_f = 10$, for example, the contaminant is predicted to move at 1/10th the speed of the carrying fluid. The latter is determined by the methods outlined in Chapter 4. Slow-moving dissolved contaminants have high retardation factors. Rapidly moving contaminants have low retardation factors. The retardation equation assumes equilibrium between the sorbate in solution, and the solid sorbent, at the pore or fracture scale. That is, it assumes the contaminants adsorb, desorb, and diffuse into intergranular porosity instantaneously. It also assumes that only a small fraction of the available sites will ever be occupied by the contaminant. The equation thus predicts that if a plume leaks from a tank and through a soil column, any sorbed contaminant will desorb and move back into the biosphere as soon as the source is removed and fresh water comes into contact with soil surfaces. The retardation equation is derived from very simple geometrical arguments, and the "retardation" is due purely to the fact that contaminants temporarily sorbed to the rock or soil have zero velocity. Figure 5.8 makes the point that in most cases order of magnitude differences in K_d's cause the same order differences in the distance a contaminant can travel in a set amount of time.

K_d's are useful in an order of magnitude sense, yet they lump together a large number of microscopic processes that may differ in detail between contaminant-soil pairs (Brady and Zachara, 1996; Davis and Kent, 1990). Consequently, K_d's are not completely portable, though they do give a rough ideal of the affinity of a given contaminant for a particular soil type. In soils hydrophobic organic contaminants sorb most strongly to soil organic matter, if the latter is available. The K_d describing sorption for a wide variety of organic molecules is a linear function of the octanol-water coefficient, mentioned above, and the fraction of organic matter in the soil

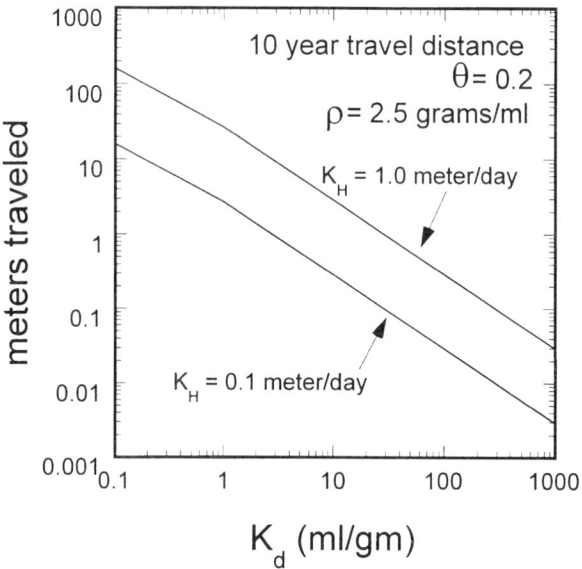

FIGURE 5.8 Ten-year travel distance calculated from Equation 5.1 as a function of K_d for $K_h = 0.1$ and 10.0 meter/day at $\Delta h/\Delta l = 0.1$; porosity = 20%; soil density = 2.5 grams/ml.

(e.g., Karickhoff et al., 1979). There exists a critical soil organic content below which organic compounds primarily attach to mineral surfaces. This critical level is proportional to the amount of available mineral surface area but also depends on the K_{ow} of the compound. K_{ow} is the ratio of a given chemical present in octanol vs. the amount of the same chemical in water (see McCarty et al., 1981).

In fact many contaminants sorbed for any significant length of time attach irreversibly, become over-coated and trapped, or diffuse into dead-end, intergranular porosity (see Figures 5.2 and 5.3). The combined effects constitute a net, and often major, reduction in soil toxicity, as contaminants thus sequestered are largely isolated from the biosphere. To the extent that radionuclides decay and other such-sorbed contaminants are prevented from escaping, they are "passively remediated" (see below).

Metal release from soil minerals is specific to metal-soil pairs and reflects the atomic interactions occurring at the mineral-solution interface. It is common to group the atomic-level interactions into adsorption and desorption rate constants k_a and k_d: (e.g., Adamson, 1990)

$$dS/dt = k_a \cdot C \cdot (S_0 - S) - k_d \cdot S$$

for the Langmuir model, and

$$dS/dt = k_a \cdot C^n - k_d \cdot S^m$$

for the Freundlich model, where S is the number of moles of metal sorbed per unit surface area of the solid substrate, S_0 is S at saturation (related to the maximum number of available sites per area), C is the metal concentration in solution, and n and m are exponents determined by experiment.

At steady-state, $dS/dt = 0$, and metal concentrations in solution are related to sorbed concentrations by an "isotherm"; then:

$$C = [(k_a/k_d) \cdot (S_0/S - 1)]^{-1}$$

for a Langmuir isotherm, and

$$C = (k_a/k_d)^n \cdot S^{n/m}$$

for a Freundlich isotherm. The common linear K_d model is a special case of the Freundlich for $n/m = 1$, and $K_d = (k_a/k_d)^n$. The isotherms are usually measured in adsorption experiments, where the aqueous solution was given an initial concentration C_0 of the metal, and the final solution concentration at the end is used to infer S. Very few workers have measured desorption isotherms, where the solid is pre-saturated with the metal, then placed in a solution with $C_0 = 0$; the solution C climbs, and reaches some equilibrium value which is again used to infer S by mass balance. The solution is then removed, and new solution introduced with $C_0 = 0$, and the process repeated until an isotherm plot of C vs. S is obtained.

If the adsorption/desorption process is reversible, and adequate time given to reach equilibrium, the adsorption and desorption isotherms should be identical. In reality, the shapes of adsorption and desorption isotherms are often quite different (e.g., Barney, 1984). For many radionuclides the K_d inferred from desorption experiments is much higher than that inferred from adsorption experiments (Adsorption K_d's are measured by adding a metal salt to a mineral solution slurry, withdrawing the supernatant, and analyzing the latter for dissolved metal levels, and subsequently calculating a K_d by difference). There are several probable causes for the discrepancy, including a simple failure to allow adequate time for equilibrium in the desorption process, since desorption tends to be much slower than adsorption ($k_d \ll k_a$); and chemical and physical changes in the mineral surface with time. Formation of a coating phase in which the trace metal (or organic molecule) is a major constituent will tend to buffer solution concentrations until that phase is exhausted, so the C(S) isotherm will simply be a line with C = constant.

If the contaminant source were removed, and metals were to desorb rapidly, the level in solution would reflect equilibrium, and the net flux from soil mineral to soil solution would be determined by the rate at which fresh fluids moved past the surface. Rapid desorption is a limiting case which rarely describes the release of metals from actual soils, pointing to metal stabilization, or occlusion, at the mineral-solution interface, or simply a $1/k_d$ that is long compared to the characteristic advection time through the system.

Historically, the study of desorption phenomenon has taken a back seat to the unraveling of adsorption behavior. Adsorption isotherms describing metal uptake by surfaces in contact with metal-rich solutions help constrain the net retardation of contaminant plumes (or potential leakage from nuclear waste repositories) by aquifer/soil minerals. Desorption isotherms and rates are needed to understand a different, and arguably more pressing problem, namely the long-term persistence and health impacts of contaminant plumes after the source has been removed. Apparently, many sorbed metals are strongly or irreversibly sorbed, desorbing very slowly if at all. This fraction is, depending on the desorption rate, unavailable to the biosphere. Engineered removal of this fraction is unnecessary if it can be demonstrated to be immobile.

The sorption of contaminants onto soil minerals often constrains the choice of contaminant removal techniques. The best example of this is the application of pump-and-treat methodologies to remove sorbed organics. Pumping and treating simply involves the pumped removal of contaminated water from a contaminant plume, treatment of the contaminant at the surface and disposal of the treated water to surface drainage, storm sewers, or reinjection back into the aquifer. When faced with clean water, reinjected or provided by natural recharge, the contaminant may desorb from soil surfaces, recontaminating groundwater (Cole, 1994; Freeze and Cherry, 1979; Isherwood et al., 1993). This means that multiple flushing is required, at least until all of the once-sorbed contaminant has been removed, and/or contaminant levels have reached regulatory targets. In practice hundreds of rinses may be required (Bear et al., 1994; Rice et al., 1995) which points to remediation times on the order of decades. Remediation effectiveness and duration depend on the contaminant mass, extent of adsorption, rate of desorption, and the time and effort required to move multiple volumes of water through the aquifer. For contaminant plumes in fine-grained soils having high surface areas and low hydraulic conductivities, the extent to which a soil can be treated by flushing in a timely fashion is limited. Because of the accumulated difficulties pump-and-treat methodologies rarely reach their cleanup goals (see below).

BACKGROUND READING

Bethke C. M. (1996) *Geochemical Reaction Modeling.* Oxford University Press, New York.
Brady P. V. and J. M. Zachara (1996) Geochemical Applications of Mineral Surface Science, in *Physics and Chemistry of Mineral Surfaces,* P.V. Brady, Ed., CRC Press, Boca Raton, FL.
Dragun J. (1988) *The Soil Chemistry of Hazardous Materials,* Hazardous Materials Control Research Institute, Silver Spring, MD.
Morel F. M. M. and Hering J. (1993) *Principles and Applications of Aquatic Chemistry.* Wiley-Interscience, New York.
Pankow J. (1991) *Aquatic Chemistry Concepts.* Lewis Publishers, Boca Raton, FL.
Rai, D. and Zachara, J. M. (1984) *Chemical Attenuation Rates, Coefficients, and Constants in Leachate Migration.* Electric Power Research Institute, Palo Alto, CA.
Stumm W. and Morgan J. J. (1996) *Aquatic Chemistry.* Wiley Interscience, New York.

6 Biodegradation

FATE AND TRANSPORT

Nearly 800 billion gallons of petroleum hydrocarbons are shipped around the globe per year — it is no surprise that some of it spills and contaminates soils (Chapelle, 1993). At the same time vast quantities of hydrocarbons are refined and turned into a daunting number of products, from pesticides to plastics. As with natural biomass, much of this material is broken down by organisms, or by the sun, ultimately into carbon dioxide and water. The breakdown occurs in the atmosphere, in soils, and in natural waters. Degradation rates are drastically different depending on where breakdown occurs, and by what means. Consequently it is critical that the fate and transport of the various organic molecules be understood. Fortunately, there exists a fairly useful, and relatively simple, set of methods by which the pathways can be worked out (e.g., Ney, 1995). There are six things that can happen to organic contaminants.

1. They can be photolysed (in essence, sunlight pushes an organic contaminant down the path to carbon dioxide).
2. They can be degraded by microbes (see below), plants, and/or animals.
3. They can be hydrolysed by water and its constituents.
4. They can be diluted and dispersed to insignificance.
5. They can be sorbed to minerals, colloids, organic matter disseminated in soils.
6. They can accumulate in plants and animals.

Obviously, if the first five processes occur to any great extent, the sixth might be avoided. It turns out that the extent of each process can be roughly characterized by two chemical characteristics — solubility in water and octanol-water partitioning. As outlined earlier the solubility of a substance is the degree to which it can be maintained in the dissolved state. The solubility of a substance gives important clues about the way it behaves in a number of other processes, and consequently, says a lot about where it ultimately ends up in the environment. The higher the solubility of a contaminant, the higher its mobility, and the less likely it is to accumulate in soils and organisms, and persist in the environment. If the solubility of a contaminant in water is low, it is more likely to stick to minerals, stay out of groundwater, and, if ingested by an organism, remain there.

Ney (1995) groups organic contaminants into three classes: low solubility — solubility less than 10 ppm; high solubility — solubility greater than 1,000 ppm; and medium solubility — between 10 and 1000 ppm. In order of decreasing solubility organic contaminants follow the rough sequence: halogenated C1 and C2 compounds > alkylated benzenes > chlorinated benzenes > PCB's. PAH's and linear alkanes cover a wide spectrum of solubilities. Toluene (one benzene with one methyl group) has

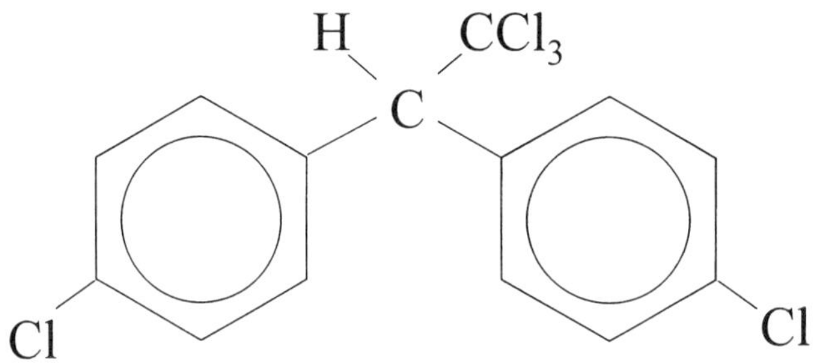

FIGURE 6.1 DDT.

a high solubility of 1780 mg/L. Pyrene (four attached benzene rings) has a low solubility of 0.135 mg/L. Some of the more soluble alkane and alkene contaminants include: chloroform (8000 mg/L), carbon tetrachloride (800 mg/L at 20°C), and trichloroethene (1100 mg/L at 20°C) (Fetter, 1992).

K_{ow} is measured in the lab, and is found to give a rough measure of the potential for bioaccumulation. Contaminants with high K_{ow}'s are more likely to sorb to organic matter in soils and to accumulate in the food chain, and vice versa. Ney (1995) groups these into three classes as well: low K_{ow} contaminants have an octanol-water coefficient less than 500; high K_{ow} contaminants have an octanol-water coefficient greater than 1000. Low K_{ow} contaminants tend to be water-soluble, mobile, less prone to bioaccumulation, and more likely to be biodegraded. High K_{ow} contaminants aren't generally water-soluble, and tend to adsorb, as well as resist biodegradation. Contaminants having mid-range K_{ow}'s can behave in either fashion. Useful compilations and explanations regarding K_{ow}'s and solubilities can be found in Kenaga and Goring (1980) and Lyman (1982). Two examples, DDT (see Figure 6.1) and benzene, serve to illustrate the means for estimating environmental pathways of contaminants. The solubility of the pesticide DDT is 0.0017 ppm. Its K_{ow} is 96,000. By the criteria stated in the paragraphs above we would predict that DDT will remain in or on soils. It should bioaccumulate, and it should be biodegraded very sparingly. The primary means of exposure is, therefore, likely to be through the consumption of contaminated food, as opposed to the consumption of contaminated groundwater. This is, in fact, the behavior which is observed. Benzene has a solubility of 1780 ppm and a K_{ow} of 83. We should expect benzene to be mobile in groundwaters, poorly sorbed, and rapidly biodegraded, which is in fact what is observed. More elaborate approaches exist for quantifying fate and transport (e.g., Hamaker, 1979; Ney, 1995).

Because chemicals which are least soluble are also most likely to sorb, there should be a correlation between solubility and K_d; and in fact, for organic molecules, there is. Kenaga (1980) relates the K_d for adsorption of organic molecules (K_{oc}) to the organic fraction of soils (f_{oc}) as:

$$\log K_d = \log K_{oc} + \log f_{oc} = 3.64 - .55 \log S + \log f_{oc}$$

where S is the solubility of the organic molecule in mg/L. If we assume a ball-park organic fraction in soils of 1% (0.01) then this means that if the solubility of an

Biodegradation

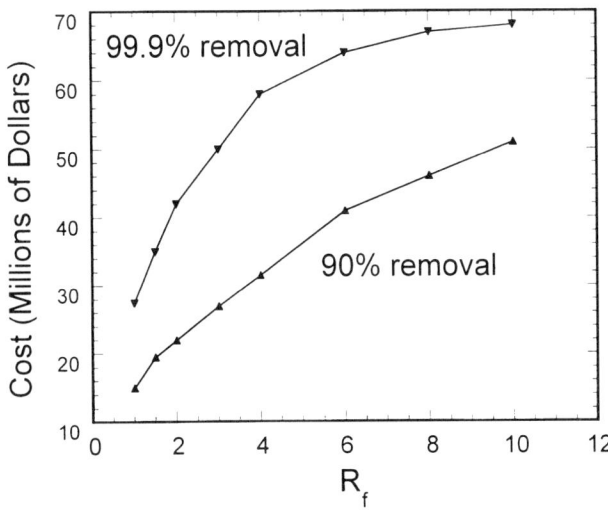

FIGURE 6.2 Cost of operating a pump-and-treat system as a function of a contaminant's retardation factor. The example assumes a pumping rate of one pore volume per year, initial capital costs of $650,000, initial operation and maintenance costs of $180,000, a 3.5% discount rate (reflecting 7.4 and 4% interest and inflation rates), and complete replacement of equipment every 25 years (Used with permission from National Research Council, 1994b, National Academy of Sciences, Courtesy of the National Academy Press, Washington, D.C.)

organic contaminants is less than about 400 mg/L its K_d will be greater than 2 mL/gm (for example, if S = 1000 mg/L, K_d = 1 mg/L; if S = 10 ppm, K_d = 13 mg/L; if S = 1 ppm, K_d = 44 mg/L). For most important organic contaminants, value of K_{oc} can be downloaded readily from:

http://www.epa.gov/superfund/oerr/soil/index.html#fact (get attachment C).

Sorbed contaminants are an important obstacle to pump-and-treat methods for removing contaminants. Contaminants which are sorbed appreciably require much higher volumes of soil flushing solutions, which makes cleanup times much longer. Figure 6.2 makes the point that a R_f much greater than 3 or 4 can drastically increase (i.e., double) the cost of cleanup. For most soils this is equivalent to a K_d of around 0.3. There are very few contaminant-soil pairs which have K_d's this low. They are almost always higher, which means that the cost of cleanup is likely to be multiplied to an even greater cost since it is predicated on contaminant removal by repeated flushing. Note also that slow desorption can complicate matters further. There have been cases where the easily accessible contaminants were removed, and the water made reasonably clean, only to have slow desorption of pollutants recontaminate the aquifer. Although the cost escalation will be different from site-to-site, the point remains — remediation gets very expensive when contaminants sorb.

In soils the primary sink for organic contaminants is generally breakdown by soil microorganisms. These come in four types: procaryotes (bacteria which don't have a true nucleus), eucaryotes (algae, fungi, and protozoa), archaebacteria (primitive methane producers), and viruses. Procaryotes and archaebacteria play the primary role in the breakdown of organic debris in soils. To varying degrees, soil microbiota use organic molecules as a source of energy and carbon. When microbiota consume organic molecules they must pass the electrons gained to an electron acceptor. Aerobic degradation involves oxygen as the electron acceptor. Anaerobic degradation involves the passing of electrons to an acceptor other than oxygen, the primary ones being NO_3^-, NO_2^-, Fe^{3+}, Mn^{4+}, SO_4^{2-}, and ultimately CO_2. Degradation rates of xenobiotic contaminants depend largely on the electron acceptor identity and availability and the inherent reactivity of the organic molecule, and/or substrate availability (Rittman and McCarty, 1980).

Before examining the pathways by which microbiota break down xenobiotic (man-made) organic molecules, the abiotic degradation of the latter must be dealt with. Hydrolysis is an abiotic reaction which can only happen to organic molecules possessing hydrolyzable groups: esters, aliphatic halogens, amides, carbamates and phosphate esters. Hydrolysis involves the attack and transformation of one of the above functional groups by H_2O, H^+, or OH^-. Hydrolysis half-lives have been tabulated by Howard (1987) and can vary by orders of magnitude depending on a variety of soil chemical factors (Perdue, 1983).

If a soil system is closed (i.e., the net import of material from the atmosphere and/or external fluids is minimal) there is a fairly predictable sequence of electron acceptor utilization which reflects the relative ease at which the specific molecules involved acquire electrons. First, O_2 is exhausted, followed by nitrates and nitrites, Mn^{4+}, Fe^{3+}, and sulfate. CO_2 is early on converted to organic acids, but is primarily used after the available sulfate is consumed to produce methane, CH_4. The relevant reactions are shown in Table 6.1, using benzene, C_6H_6, as the organic contaminant (electron donor), and schematically in Figure 6.2. Different bacteria facilitate different portions of the overall degradation path shown in Figure 6.3. In soils the appropriate bacteria appear to be widely available (e.g., Ludvigsen et al., 1995; Stumm and Morgan, 1996). Moreover, there is substantial evidence that microorganisms adapt to take advantage of the available electron acceptors to biodegrade organic contaminants more effectively (e.g., Chiang et al., 1989). Consequently it is safe to assume in most cases that the trend in Figure 6.3 will not be limited by whether or not the right microbe exists in the soil. For the case of BTEX degradation there seems to be no particular correlation between microbial cell numbers and BTEX levels, or for that matter SO_4^{2-}, HCO_3^- and H_2S (Toze et al., 1995).

Generally when dissolved O_2 levels in soil solutions are 2 mg/L or higher (Salanitro, 1993) and/or gaseous O_2 in the vadose zone is greater than 5% (Atmosphere O_2 is around 20%) oxygen is the primary electron acceptor (Brown et al., 1995). If the rate of organic carbon degradation exceeds the supply of oxygen from the atmosphere and soil zone, nitrate and/or nitrite, become the primary electron acceptors. Rates of oxygen replenishment are diminished when soils are tightly compacted and/or water-logged.

Nitrate in soils comes from a variety of sources and is generally produced by oxidation of NH_4^+. The primary sources of the latter in soils are fertilizers, animal wastes and the conversion of atmospheric N_2 by specific types of soil bacteria (N_2

TABLE 6.1
Benzene Degradation with Various Electron Acceptors

Degradation	Electron donor	Reaction	Electron source
Aerobic	O_2	$C_6H_6 + 7.5O_2 \Rightarrow 6CO_2 + 3H_2O$	Atmosphere
Denitrification	NO_3^-	$C_6H_6 + H^+ + 6NO_3^- \Rightarrow 6CO_2 + 3N_2 + 6H_2O$	Agriculture, atmosphere
Iron-reduction	$Fe(OH)_3$	$C_6H_6 + 30Fe^{3+} + 12H_2O \Rightarrow 6CO_2 + 30H^+ + 30Fe^{2+}$	Fe-(hydr)oxides
Sulfate-reduction	SO_4^{2-}	$C_6H_6 + 7.5H^+ + 3.75SO_4^{2-} \Rightarrow 6CO_2 + 3.75H_2S + 3H_2O$	$CaSO_4$
Mn-reduction	MnO_2	$C_6H_6 + 12H_2O + 15Mn^{4+} \Rightarrow 6CO_2 + 15Mn^{2+} + 30H^+$	Mn-(hydr)oxides
Methanogenesis	CO_2	$C_6H_6 + 4.5H_2O \Rightarrow 2.25CO_2 + 3.75CH_4$	Atmosphere, minerals

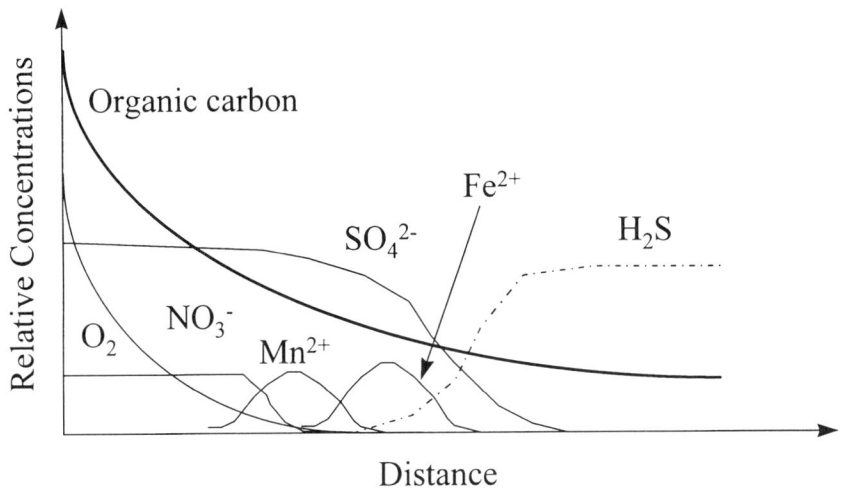

FIGURE 6.3 Schematic of closed-system evolution of electron donors caused by degradation of an organic contaminant plume as a function of distance. (From Stumm, W. and Morgan, J. J., *Aquatic Chemistry,* Wiley-Interscience. With permission of John Wiley & Sons, Inc., New York.)

fixation). As with oxygen, when the degradation rate exceeds the rate of NO_3^- replenishment, a net drawdown of the latter occurs and, ultimately, the electron donor capacity arising from soil nitrate is exhausted. At this point further degradation of organic carbon depends for the most part on the availability of oxidized iron, sulfate, and manganese.

Oxidized iron (Fe^{3+}) is present in many soils as poorly crystalline (hydr)oxides, where they are potential electron acceptors for the oxidation of organic contaminants. The reaction produces highly soluble Fe^{2+}, and the presence of the latter can be used to track utilization of Fe^{3+} as the primary electron acceptor for degradation. In the absence of ongoing reduction Fe^{3+} is otherwise very insoluble and present in solution at only trace levels. Indeed groundwaters enriched in Fe^{2+} by degradation of organics generally drop their iron once they come into contact with oxygen.

Manganese can exist in three valence states in soils and soil waters; +2, +3, and +4, and can form mixed-valence (hydr)oxides. At the same time trivalent Mn can disproportionate into divalent and quadrivalent Mn. In natural waters Mn exists primarily as Mn^{2+}. The most important Mn-(hydr)oxides, MnOOH, MnO_2, and Mn_3O_4, contain Mn primarily in the higher two valence states (for extensive reviews of Mn chemistry see Hem (1989) or Stumm and Morgan (1996). There is an extensive literature attesting to the reductive dissolution of Mn (hydr)oxides by organic matter (see (Deng and Stone, 1996; Stone, 1987), whereby the oxidized Mn centers acquire electrons from adsorbed organic molecules, which are themselves transformed to more oxidized daughters. Sulfate in soils and groundwaters comes from dissolution of sulfur-containing minerals, sulfate-containing rainfall, and seawater intrusion in some cases. The availability of CO_2 as an electron acceptor depends upon its supply by biologic respiration, diffusion from the atmosphere, dissolution of carbonate minerals, and decay of organic matter elsewhere.

The theoretical efficiencies for biodegradation of BTEX compounds are fairly well worked out: 1.0 mg/L of dissolved oxygen, when used by microorganisms, can degrade 0.32 mg/L of BTEX; 1.0 mg/L of dissolved nitrate degrades 0.21 mg/L (Wiedemeier et al., 1995b). Each mg/L of degraded BTEX should result in the production of 21.8 mg/L of Fe^{2+}, and each mg/L of sulfate should result in the degradation of 0.21 mg/L of BTEX (Wiedemeier et al., 1995b). For an extensive discussion of the free energies driving the individual biodegradation reactions, and examples where reaction trends are calculated see Wiedemeier et al. (1995b).

Given available electron acceptors, and nutrients, biodegradation can occur by any of a number of pathways. Organisms use enzymes to catalyze, that is lower the energy needed to convert organic compounds for metabolism, and thereby dramatically enhance degradation rates. Although many organic contaminants, because of their engineered structures, are somewhat "new" to many enzymes, it appears that most organisms possess a number of enzymes sufficiently versatile to overcome the initial unfamiliarity. As pointed out by Schwarzenbach et al. (1993) because organisms have long been bombarded by chemicals from other species, they have always had the need to eliminate some of the chemical noise, hence the non-specific enzymes. Organisms convert xenobiotic compounds to more familiar, easier to metabolize forms by either using biochemical energy to make aggressive reacting species, or by catalyzing degradation along hydrolytic pathways (Schwarzenbach et al., 1993).

Aerobic degradation of linear alkanes generally involves the oxidation of terminal methyl groups to carboxylates, which are subsequently oxidized to CO_2 plus water (see Figure 6.4). Aromatics are generally broken down under aerobic conditions through cleavage of the ring by molecular oxygen, and subsequent formation of two adjacent OH groups, which then form a dicarboxylic acid. Polycyclic aromatics are similarly broken down under aerobic conditions by O_2 cleavage of constituent rings to form carboxylates (see Figure 6.5). Anaerobic degradation of many aromatic contaminants occurs by oxidation to phenols or organic acids, then long-chain

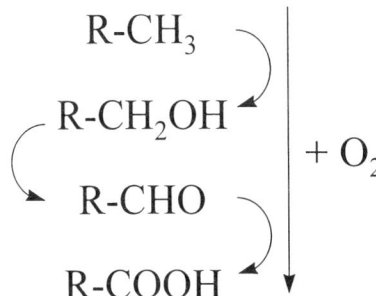

FIGURE 6.4 Aerobic degradation of terminal methyl group. (After Chapelle, 1993).

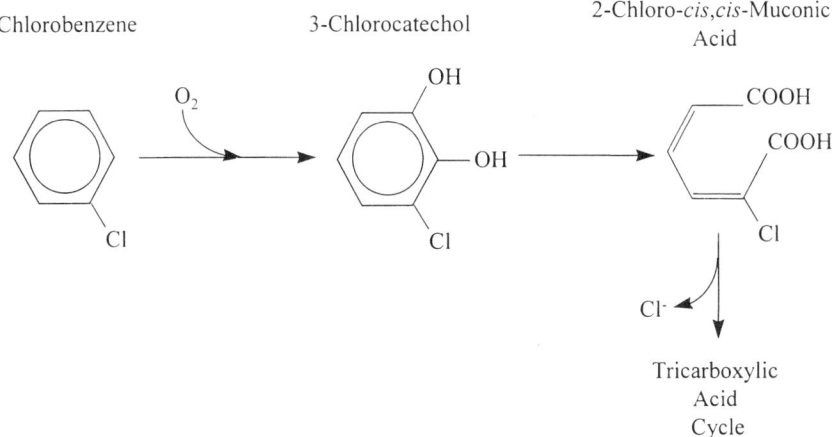

FIGURE 6.5 Aerobic degradation of aromatic. (After Chapelle, 1993).

volatile fatty acids, which are then transformed to methane and carbon dioxide (Grbic-Galic, 1990).

Chlorinated organics, such as PCE and TCE, are relatively oxidized to begin with. Consequently, there is less energy to be gained by a microorganism in oxidizing them further. Instead, they are used as electron acceptors and are reduced to less oxidized, and less chlorinated daughters. This process, which is the most significant degradation pathway for highly clorinated solvents (Wiedemeier et al., 1996), is called reductive dehalogenation, and involves the exchange of hydrogen for an attached halogen. The sequence of reductive dechlorination of tetrachloroethylene produces TCE, then dichloroethylene, vinyl chloride, and ultimately ethylene, ethane, and methane (Vogel and McCarty, 1985) — see Figure 6.6. The less chlorinated daughters are, in turn, more reduced, and resistant to further reduction. Vinyl chloride, with one chlorine, degrades particularly slowly in reducing environments. On the other hand, if transported into an oxygen-rich zone, vinyl chloride (and the other chlorine-poor

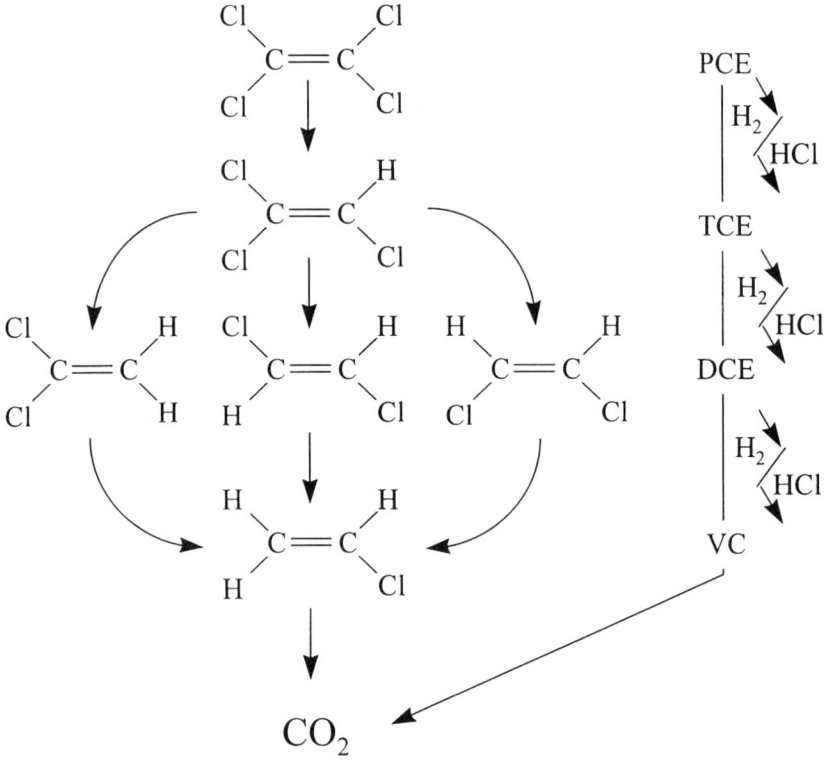

FIGURE 6.6 Sequential breakdown of PCE. (After Chapelle, 1993).

daughter products) is rapidly oxidized, ultimately to CO_2 and water. Figure 6.7 shows breakdown paths for PCE, TCE, TCA, and carbon tetrachloride.

In essence, the transition from PCE (or TCE) to CO_2 (or methane) is constrained by the redox state at the front and the back end, and both must be different for the whole sequence to work rapidly. Breakdown of PCE is most prevalent when conditions are reducing, and reductive dechlorination is most complete when conditions are most reducing — methanogenic conditions are better than sulfate or nitrate-reducing conditions. Breakdown of vinyl chloride, however, requires oxidizing conditions. (There are exceptions — Hale et al. (1996) have observed appreciable TCE breakdown under oxygen-rich conditions.) In effect, chlorinated organics act as electron acceptors at the front end, and electron donors at the back end. For the reactions to occur, therefore, there must be ample electron donors near the contaminant source, and abundant electron acceptors away from the source. One effect of this need for biomodal redox conditions is that demonstration of natural attenuation of chlorinated organics often requires a clear determination of the spatial and temporal distribution of redox conditions (see below).

Co-metabolic breakdown of chlorinated organics is contingent on an organism metabolizing something else, such as methane or toluene. The microorganisms gain

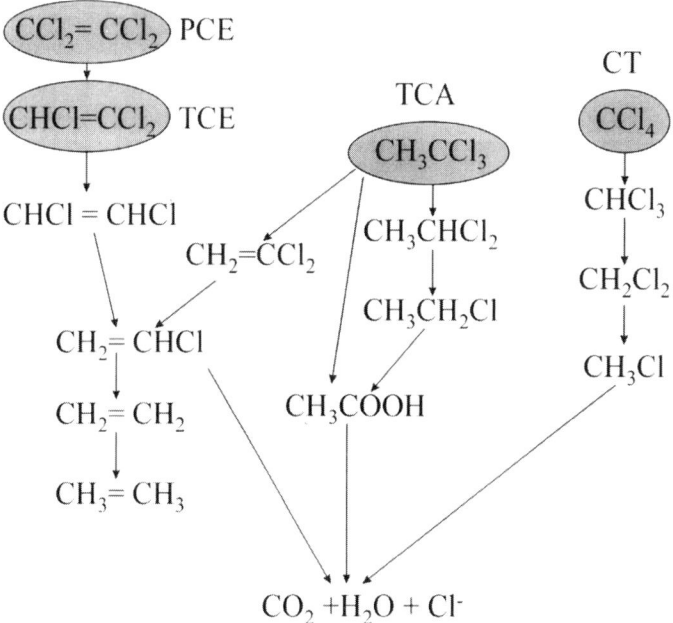

FIGURE 6.7 Degradation paths for PCE, TCE, TCA, and CT. (After McCarty, 1994).

no energy from the breakdown of the chlorinated organic, but from the breakdown of the non-chlorinated substrate.

Biodegradation of organic contaminants is a complex process, and the finer points of the specific pathways are still being worked out. A number of excellent sources detail the evolving state-of -the art. Vogel et al. (1987) provides an extensive listing of half-lives for halogenated hydrocarbons, as well as an exhaustive review of metabolic and abiotic degradation pathways. Thorough coverage of the kinetics of organic reactions and microbial growth is presented by Schwarzenbach et al. (1993). An exhaustive examination of soil microbiology is that of Chapelle (1993), and an extensive examination of the microbiological pathways for BTEX degradation is that of Wiedemeier et al. (1995a).

For our purposes, an understanding of the specific degradation pathways is less important than an awareness of the overall trends. Extensive compilations of degradation half-lives (e.g., Howard et al., 1991) demonstrate that degradation of xenobiotic contaminants in soils is more the rule than the exception. In 1993 the National Research Council (National Research Council, 1993) examined *in situ* bioremediation and classified the biodegradability of various organics as either: "established" or "emerging". These are listed in Table 6.2.

Degradation is generally faster and more complete when oxygen is the electron acceptor. Also, chlorinated hydrocarbons generally degrade slower than non-chlorinated hydrocarbons. Biodegradation of poorly-sorbed molecules is faster than degradation of strongly sorbed organics. Biodegradation of fuel hydrocarbons proceeds

TABLE 6.2
Contaminant Susceptibility to Bioremediation

	Organic	Notes
Established	Hydrocarbons and derivatives (gasoline and fuel oil)	Relatively rapid aerobic and anaerobic degradation
	Alcohols + ketones + esters	
Emerging	PAHs	Aerobic degradation in some cases
	Creosote	Readily degradable under aerobic conditions
	Ethers	Degradable under aerobic or nitrate-reducing conditions
	Highly chlorinated aliphatics	Cometabolized under some conditions
	Less chlorinated aliphatics	Aerobic degradation; cometabolized by anaerobic microbes
	Highly chlorinated aromatics	Aerobic degradation; cometabolized by anaerobic microbes
	Less chlorinated aliphatics	Aerobic degradation
	Highly chlorinated PCBs	Cometabolized by anaerobic microbes
	Less chlorinated PCBs	Aerobic biodegradation
	Nitroaromatics	Anaerobic and aerobic degradation

From National Research Council, 1993.

along the trend: linear/branched alkanes + monocyclic saturates > tricyclic saturates > dicyclic saturates >> pentacyclics/tetracyclics (Huesemann, 1995). The identity of the molecule may therefore be a more important determinant of its degradation rate than soil type, fertilizer levels, or microbe count (Huesemann, 1995). Degradation of linear alkanes is for the most part faster than degradation of branched alkanes. In water-dominated systems (e.g., streams) biodegradation is largely controlled by sediment-dwelling bacteria (see e.g., Cohen et al., 1995). Although degradation is often assumed to be slower in colder climates, recent toluene degradation studies (Bradley and Chappelle, 1995) suggest that there may be exceptions to this rule. Last, the accessibility of an organic contaminant to organisms is important. Contaminants sequestered in dead-end pores are sometimes unavailable for uptake (see Chapter 8).

Table 6.3 is a compilation of field and soil microcosm studies demonstrating relatively rapid biodegradation of organic contaminants. The studies cover degradation of a variety of organic contaminants degraded under aerobic, nitrate-reducing, iron-reducing, sulfate-reducing, and methanogenic conditions. The list is by no means exhaustive, but rather, is meant to emphasize that there is a wealth of scientific evidence attesting to organic degradation of many contaminant molecules.

Table 6.4 summarizes soil half-lives for a number of organic contaminants in soils (Howard et al., 1991) which are useful for comparison. The data combine both hydrolysis and biodegradation and are based on a number of assumptions. The most important of these are

1. Some of the soil half-lives are based on water half-lives (this is particularly true for hydrolysis half-lives). Some of the latter were generated from screening studies and/or empirical correlations (Howard et al., 1987).

TABLE 6.3
Recent Studies Demonstrating Significant Biodegradation

Contaminant	Reference
BTEX, Aviation fuel and individual components	(Barker et al., 1987; Bradley and Chappelle, 1995; Chiang et al., 1989; Cohen et al., 1995; Hadley and Armstong, 1991; Kim et al., 1995; Lovley et al., 1989) and several others
Chlorinated ethanes and ethenes (TCE, PCE, VC, etc.)	(Cline and Viste, 1985; Davis and Carpenter, 1990; Fetter, 1989; Hopkins et al., 1993; Nelson et al., 1988) and others.
Phenols and/or PAH's	(Erlich et al., 1982; Godsy et al., 1992; Klecka et al., 1990; Madsen et al., 1991; Millette et al., 1995; Nielsen et al., 1996; Wilson et al., 1985)
PCB's	(Fish, 1996; Olson,; Otfjord et al., 1994)
Others: CFC-113 (Freon®), PCDD/PCDF (penta- to heptachlorinated dibenzo-p-dioxin and dibenzofuran), Nitroaromatic compounds	(Adriaens et al., 1995; Heijman et al., 1995; Jackson et al., 1990)

Where groundwater data were absent, biodegradation half-lives were often set equal to twice the surface water values on the assumption that lower microbial populations and enzymatic capability, and the lesser likelihood of aerobic conditions, would decrease rates.

2. The range of half-lives given for each chemical represents the highest high and the highest low, given two competing reactions (e.g., hydrolysis and biodegradation).

Figure 6.8 illustrates schematically the decay in contaminant levels as a function of time and degradation half-life.

Comparison of the half-lives from Table 6.4 with the schematic in Figure 6.8 shows that, with the exception of DDT, and potentially TCE and PCE, the other organic molecules degrade very rapidly on a ten-year-or-less time frame. Table 6.4 obviously represents a small fraction of organic contaminants. Although the listed half-lives are typical of organic contaminants and soil carcinogens, many of which have soil half-lives of 2.5 years or less (Borgert et al., 1995), the numbers in Table 6.4 say nothing about soil-biota-contaminant interactions which may be site and/or species specific. The complexity of these combined factors assures that contaminant half-lives measured at one site are not completely portable to other sites. The degradation measurements, taken as a group, indicate that the large majority of organic contaminants degrade in soils on a human time scale without human intervention, and that it is overly conservative to assume that degradation won't occur. This being said, for the reasons cited above the rates in Table 6.4 should only be used in contaminant transport models as a last resort. Their indiscriminate application to sites where biodegradation is not occurring could lead to serious errors.

TABLE 6.4
Degradation Half-Lives Of Selected Organic Contaminants

Contaminant	Half-life range[a]
DDT	2–15.6y
insecticide — banned in the U.S. in 1972	16d–31.3y
	16–100d
Carbon tetrachloride	6–12m
solvent	7–365d
	7–28d
Benzene	5–16d
gasoline, manufacturing	10d–24m
	16w–24m
Methyl bromide	7d–4w
fumigant	14–38d
	28d–16w
Trichloroethylene (TCE)	6m–1y
dry cleaning, metal degreasing	10.7m–4.5y
breakdown product	98d–4.5y
Tetrachloroethylene (PCE)	6m–1y
dry cleaning	1–2y
metal degreasing	98d–4.5y
Vinyl chloride	4w–6m
breakdown product	8w–95m
	16w–24m
Methyl ethyl ketone	1–7d
solvent	2–14d
	4–28d
1,2-Dichlorobenzene	4w–6m
solvent, fumigant	8w–12m
insecticide	16w–24m
Toluene	4–22d
gasoline	7d–4w
	8–30w
Cresol(s)	1h–29d
from coal-tar refining	2h–49d
	10–49d
Xylenes	1–4w
gasoline, solvents	2w–12m
	6–12m

Note: The first range is the soil value, the second is the groundwater value, and the last is the anaerobic half-life.

[a] d = days, m = months, y = years.

From Howard et al. (1991).

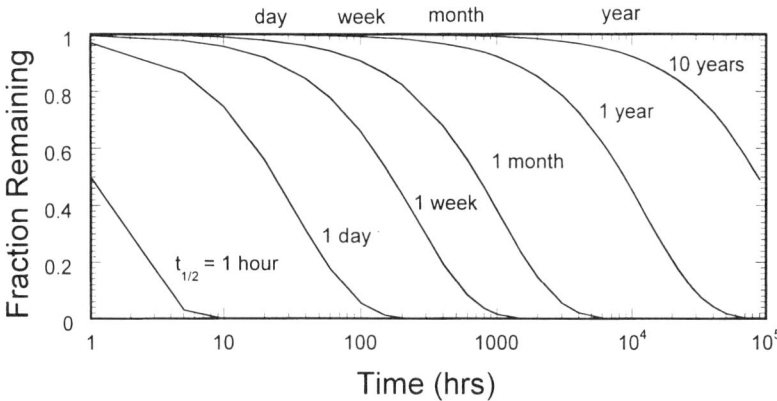

FIGURE 6.8 Residual contaminant fraction as a function of time calculated for various contaminant half-lives.

Last, it should be re-emphasized that biodegradation sometimes leads to the appearance of more toxic chemicals. At the same time a biodegradation half-life is a more changeable quantity than a radioactive half-life. While the latter depends on nuclear physics, and cannot be changed, biodegradation half-lives depend on the nature and degradative pathways utilized by the given organism(s), as well as substrate and/or nutrient availability. Since these are changeable, biodegradation half-lives are not fixed.

SUMMARY

The result of the processes outlined in Chapters 4, 5, and 6 is that contaminant plumes often slow over time, and there is an attendant drop in contaminant levels. Retarded contaminant velocities tend to arise from sorption. Lowered contaminant levels can be caused by dispersion and dilution together. In many cases the individual, or combined, effects of the two are sufficient to minimize the potential health impact of the plume. To reduce the water volume-integrated mass of contaminant, however, requires chemical reaction; precipitation/irreversible sorption and/or biodegradation. Some rules-of thumb for contaminant transport gleaned from the previous three chapters are

1. Fluid flow velocities dictate the shortest time required by a groundwater-borne contaminant to reach a receptor. However, in unsaturated soils vapor-transport of volatile contaminants may be faster than transport in solution.
2. Fluid velocities are generally higher in gravels and sands, but low in fine-trained silts and clay.
3. Subsurface heterogeneities, and the presence of NAPLs can greatly complicate and dominate contaminant transport.

4. pH's greater than neutral generally favor the removal from solution of Pb, Cd, Cu and a variety of other metals by both mineral precipitation and sorption to mineral surfaces.
5. Reducing conditions typical of organic-rich, and/or flooded soils favor the retention of Cr, Cd, and a variety of radionuclides which form insoluble solids or sorbed species in lower valence states. To the extent that Fe-hydroxides are made more soluble, reducing conditions may lead to liberation of otherwise sorbed metals.
6. Organic contaminants generally sorb to the organic fraction in soils. Sorption of organics therefore correlates strongly with the amount of soil organic matter, as well as with the polarity of functional groups.
7. Non-chlorinated hydrocarbons are rapidly degraded by soil bacteria for their electrons. The rate of biodegradation is often limited by the availability of electron acceptors (e.g., O_2, NO_3^-, SO_4^{2-}, Fe^{3+}). Aerobic degradation is typically much more rapid than anaerobic degradation.

At some point the natural capacity of a soil to degrade, sorb or precipitate a given contaminant will be exhausted due to the saturation of available surface sites, or complete utilization of the electron donors needed by organisms to degrade a particular organic contaminant, or the non-contaminant components in a precipitated solid. (Note that this depletion should not be a common occurrence so long as there is source removal.) The point at which this depletion occurs is poorly understood and must depend both on soil characteristics (e.g., electron donor availability, sorbent mineralogy, total surface area, solution composition, microflora speciation) and waste characteristics (e.g., sorbing affinity, chemical makeup, and total volume). This point will be specific to soil-contaminant pairs. Obviously, high-volume contaminants might be expected to overload the natural capacity of relatively small amounts of soil. The total capacity of a given soil to achieve oxidative degradation of organic molecules is most conveniently expressed as its oxidation capacity, OXC, which strictly monitors its ability to resist reduction (Barcelona and Holm, 1991; Scott and Morgan, 1990) and is the electron-normalized sum of the various electron acceptors (O_2, NO_3^-, Fe^{III}, Mn^{IV}, SO_4^{2-}, etc.) in a soil matrix. The analogous measure of a soil volume to resist oxidation is the reduction capacity, RDC (Heron et al., 1995), which is all of the likely electron donors (CH_4, NH_4^+, S^{-II}, S^{-I}, Fe^{II}, Mn^{II} etc.) summed on an electron equivalent basis.

The Air Force Center for Environmental Excellence (AFCEE) and the R. S. Kerr Environmental Research Center have in the past few years also sought to express the biodegradation capacity of aquifers in an easily usable form for predicting attenuation of BTEX compounds. The biodegradation capacity is one of the input parameters in the natural attenuation screening program BIOSCREEN (Newell et al., 1996) which will be covered in Chapter 8. At least for the degradation of fuel hydrocarbons there appers to be an inexhaustible supply of electron acceptors in most, if not all, hydrogeologic environments (Wiedemeier et al., 1996).

BACKGROUND READING

Borgert C. J., Roberts S. M., Harbison, R. D., and James, R. C. (1995) Influence of soil half-life on risk assessment of carcinogens. *Reg. Toxic. and Pharmacol.* (22) 143–151.

Chapelle F. H. (1993) *Groundwater Microbiology and Geochemistry.* John Wiley & Sons, New York.

Dragun J. (1988) *The Soil Chemistry of Hazardous Materials,* Hazardous Materials Control Research Institute, Silver Spring, MD.

Hem J. D. (1989) Study and interpretation of the chemical characteristics of natural water. United States Geol. Surv. Water Supply Paper 2254.

Schwarzenbach R. P., Gschwend P. M., and Imboden D. M. (1993) *Environmental Organic Chemistry.* Wiley-Interscience, New York.

Stumm W. and Morgan J. J. (1996) *Aquatic Chemistry.* Wiley-Interscience, New York.

7 Case Studies

INTRODUCTION

We propose that natural attenuation should be considered *before* selection of an engineered approach at all sites whether federal or state, CERCLA or non-CERCLA. To make this case it must be shown that:

1. The current "future" doesn't work.
2. A "future" relying on initial consideration of natural attenuation will work.

The current future is based on engineered removal of contaminant mass. For the past 26 years that has meant basically pump-and-treat. Pump and treat costs a great deal and often leaves much of the contaminant in the ground. Nevertheless, it is hard to demonstrate that newer technologies represent vast improvements.

ENGINEERED REMOVAL

High-toxicity, low-volume contaminant sources are often physically removed, whereupon the contaminated water which is left behind is treated by a variety of approaches (see Boulding, 1995; National Research Council, 1994b). Much of CERCLA-related costs have been attributed to the groundwater cleanup phase. Pump and treat is the primary remediation approach, and is used to treat contaminated groundwater at roughly three quarters of CERCLA sites (National Research Council, 1994b) despite the fact that its benefit is questionable — its poor performance at reaching cleanup goals is extensively documented (see Doty and Travis, 1991; Mackay and Cherry, 1989; Travis and Doty, 1990). Worries over the failure of pump-and-treat has led to repeated EPA reviews and an overall belief that pump-and-treat techniques can't return sites to a precontaminated state any time soon (soon being decades from now), if ever. It is difficult to demonstrate that pump-and-treat methods can ever give more than a factor of 2 or 3 decrease in dissolved contaminant levels. Note though that pump-and-treat does appear to be effective for preventing additional migration of plumes off site, particularly if applied early. The failures (and limited uses) of pump-and-treat techniques are outlined in detail in *Alternatives for Groundwater Cleanup* (National Research Council, 1994b). Newer *in situ* extraction techniques which may improve on this include (MacDonald and Kavanaugh, 1994):

- **Bioventing/*in situ* bioremediation** — Pumping air and/or nutrients through soils to stimulate organic activity.
- **Soil Vapor Extraction/Air sparging** — Pumping air through soils to remove volatile contaminants; sometimes used in conjunction with bioventing.

- ***In situ* thermal desorption/steam-enhanced extraction** — Heating to remove volatile contaminants.
- **Electrokinetics** — Electromigration of contaminants to electrodes placed in the ground.
- **Soil Flushing** — *In situ* washing of soils by injected chemicals.

A problem common to pump-and-treat, as well as the methods above, is the difficulty of getting nutrients and/or reacting chemicals into, and contaminants out of, low permeability zones (e.g., clay-rich layers). Even if there were no low-permeability zones, or other heterogeneities, at depth substantial inefficiencies arise from the nature of engineered water movement. When water (or gas) is pumped in or out from a well, there is a semi-radial delivery. By the same token, water or gas removal is radial as well. It is difficult to "aim" reacting chemicals to a given spot without putting them in a number of other spots as well. Magic bullets have less effect when the majority of their impact is directed away from their target. Note also that the soil venting and vapor extraction techniques only work well on reasonably volatile contaminants.

Soil flushing techniques involve the injection of a solubility or desorption-enhancing agent into the contaminated soil, followed by the pumped removal of the mobilizing agent and the contaminant. Smith et al. (1995) list acids (sulfuric, hydrochloric, nitric, phosphoric, or carbonic), bases (sodium hydroxide), chelating agents, reducing agents, and surfactants as potential mobilizing agents. There also remains the obstacle of getting the mobilizing agent distributed throughout the soil, and making certain that all of the contaminant mobilizing agent is ultimately removed. Note that any residual mobilizing agent might transport leftover contaminants well after the site has been considered closed. This brings up the obvious question of: Why bother spending a great deal of money (soil flushing costs run to 75 to 200 dollars per cubic yard — Smith et al., 1995) to move something out of the subsurface that left alone will remain relatively immobile and inert?

Electrokinetics is one of the few techniques which works in low permeability zones. Electrokinetics is not a new field. Nevertheless, it has only recently been demonstrated at the bench scale as a potential tool for contaminant removal from soils (Alshawabkeh and Acar, 1992; Lindgren et al., 1994; Probstein and Hicks, 1993; Runnels and Larson, 1986), and in the field, particularly through the LASAGNA™ process. The LASAGNA™ process involves the emplacement of horizontal fractures and the subsequent electrokinetic movement of contaminant-tainted fluids through reactive zones.

(See http://www.em/doe.gov/plumesfa/intech/lasagna/)

What makes contaminant extraction difficult for all of the techniques is that the longer a contaminant source exists, the more intractable (refractory) it becomes, due to diffusion into pores, and spreading. It is a fact of life that a fairly substantial amount of time elapses from identification of a problem to the implementation of remediation. It is a fact of contaminant transport that diffusion and smearing become almost unavoidable as a result. For organic molecules that sorb it appears that fast

Case Studies

paths into the soil matrix dictate their initial removal from groundwater, yet their engineered extraction is ultimately controlled by movement along slow paths (diffusion out of dead-end pores).

An increasing emphasis has subsequently been placed on more long-term, sometimes passive, solutions which focus more on the chemical reactivity, or physical presence of solids, as opposed to engineered removal of fluids. These include:

- *In situ* **Vitrification** — Relies on the fusion of soils + wastes into a glassy monolith, from which, presumably, contaminant leaching will be minimal over the long haul.
- **Reactive barriers** — Uses chemically reactive materials (e.g., zero-valent iron) to transform water-borne contaminants to less toxic reduced forms (e.g., Gilham and Burris, 1992; Tratnyek, 1996).
- **Physical containment** — Uses cutoff walls, caps, and/or liners to physically limit contaminant movement.
- **Phytoremediation** — Relies on plants to take up and metabolize contaminants at relatively shallow depths. Plants are envisioned to take up organic contaminants through transpiration, whereupon they can be oxidized, reduced, or hydrolysed; joined with other organic molecules and, ultimately, entrapped in the plant structure (see Trapp and MacFarlane, 1995).

In situ vitrification (ISV) is a technology that has been developed since 1980 by the Department of Energy to treat soils contaminated with heavy metals, and organic wastes (Dragun, 1991). In the ISV process, contaminated soil is melted *in situ* using graphite electrodes (resistance melting via Joule heating). When the target melt depth (in practice, 0 to 20 ft.) is reached, power is terminated and the melt is allowed to cool and solidify. The high temperatures attained during melting (approximately 1500°C) vaporizes water and destroys organics and combustibles. The resulting volatiles are captured by the off-gas treatment system, which consists of a hood covering the molten soil, an induced-draft fan that maintains a slight vacuum under the hood, and a filtering and scrubbing system to decontaminate the off-gases before release to the atmosphere (see Figure 7.1). Ambient air is pulled into the hood to provide cooling and an oxidizing atmosphere for combustion of pyrolysis products transported to the surface of the melt and surroundings. Non-volatile contaminants are assumed to dissolve into the melt and are incorporated into rock or glass formed upon cooling.

The ISV process has been demonstrated in large staged sites (e.g., waste placed in a subgrade engineered pits, SAIC-GeoSafe Corporation, 1995). Problems arise where significant sources of volatiles are present, such as a near surface water table or where volatile flow is inhibited by soil or rock structure. High volatilization can lead to "flashing" of groundwater, followed by a release of contaminants and slumping. At the same time, the total treatment cost for ISV is estimated to be approximately 770 dollars per cubic yard (SAIC-GeoSafe Corporation, 1995).

Reactive barriers are relatively new, so few case studies exist to demonstrate and/or calibrate long-term performance. Physical containment techniques have long been in use for municipal landfills. For both of these techniques long-term structural

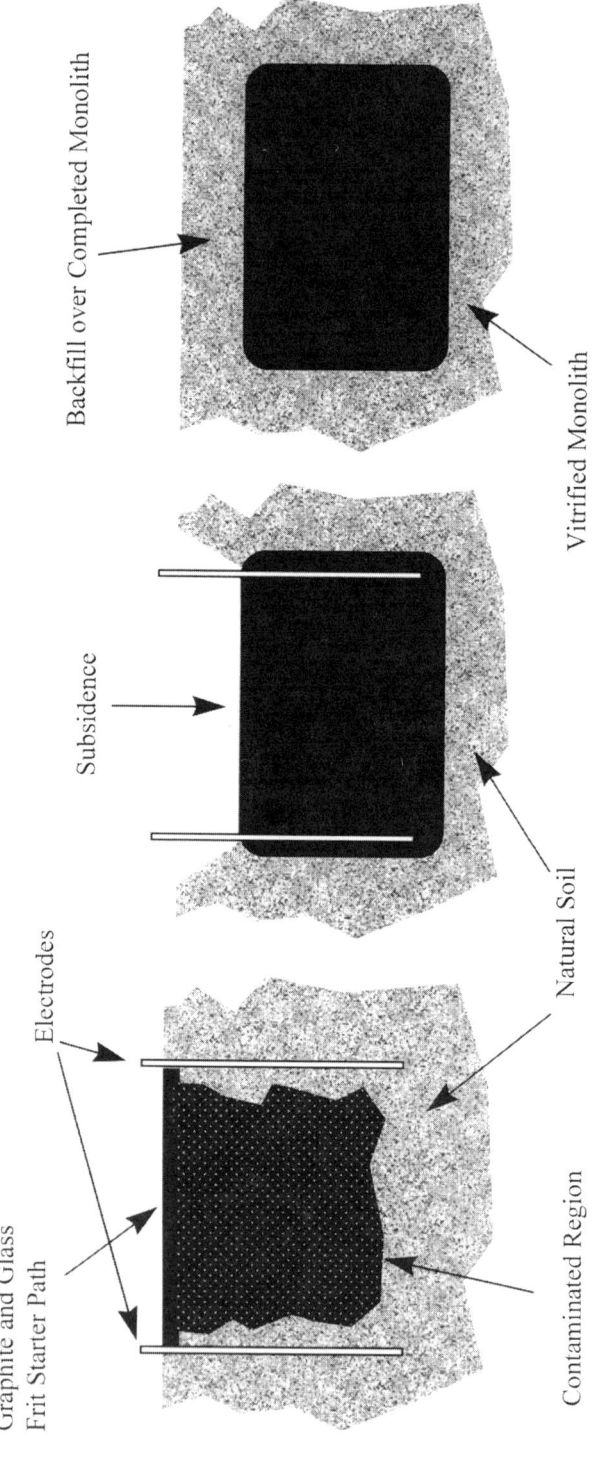

FIGURE 7.1 *In situ* vitrification.

and chemical integrity is critical. Phytoremediation is increasingly seen as an effective long-term solution at some sites (e.g., Schnoor et al., 1995).

In situ bioremediation is another solution that calls for further explanation as it is being pushed in some circles as the new magic bullet for removing contaminants, the implication being that it will do so where pump-and-treat has failed. *In situ* bioremediation is the engineered enhancement of natural contaminant biodegradation at depth. This may involve the addition of electron acceptors (e.g., O_2), or nutrients, or the organisms themselves. If the right combination of factors is brought to bear, presumably the voracity of organic activity might in the end remove the contaminants "where they live", where pump-and-treat could not. Note first of all that it is very difficult to determine *in situ* bioremediation efficiencies because of uncertainties in the amount of contaminant, the nature of the end-products, and the extent of natural attenuation occurring in the background, to name a few factors. At the same time, there are institutional obstacles to full field implementation (National Research Council, 1994b).

Many of the subsurface heterogeneities (low permeability zones with low fluid flow rates) that limit pump-and-treat also limit *in situ* bioremediation. The site conditions that favor *in situ* bioremediation include saturated hydraulic conductivities greater than 0.01 meters/second (Thomas and Ward, 1989) and a relatively uniform subsurface aquifer. Note that this hydraulic conductivity is relatively high for most soils, and corresponds to something more permeable than a silt/clay (see Chapter 4, Figure 4.1), and uniform aquifers are rare. Moreover, the high hydraulic conductivity and lack of heterogeneity needed for optimal *in situ* bioremediation are also the very narrow conditions that favor success with pump-and-treat. The very process of biodegradation can also decrease the existing hydraulic conductivities through biofouling by a factor of 1000 (Taylor et al., 1990), and, consequently, depress yields.

Soil stabilization/immobilization methods are increasingly considered for site treatment, particularly for metals. Soil immobilization techniques are designed to minimize contaminant transport by (Smith et al., 1995):

1. Reducing infiltration of fluids into the contaminant media by using barriers.
2. Reducing infiltration by modifying the permeability of the contaminant matrix.
3. Reducing the solubility, hence mobility, of the contaminant in groundwater.
4. Controlling the flow of contaminated water away from the site to permit effective collection and treatment.

One of the objects of the preceding chapters has been to demonstrate that in soils and groundwaters many of the contaminants of concern are, by their fundamental chemical nature, relatively insoluble and/or strongly sorbed to many soils, and consequently transported quite slowly.

Soil immobilization techniques include capping systems, vertical barriers, and horizontal barriers. Capping simply involves coverage of a contaminated site with a clay or synthetic membrane that is thought to be relatively impermeable. Additional effort may also be expended in grading or ditching a site to provide drainage or prevent ponding. Vertical barriers are used to prevent intrusion of uncontaminated water into, and contaminated water off of a site. Vertical barriers work best when they extend down to a horizontal, impermeable layer of rock beneath the zone of contamination (Smith et al., 1995). Slurry walls, grout curtains, sheet pile walls, and geomembrane curtains are the major types cited by Smith et al. (1995). Because slurry walls are thin, and placed in segments, the continuity of the barrier is often a major concern. Pile walls often leak between the joints. The distance of grout penetration varies for grout curtains, but can be relatively small (4.5 feet). Consequently, extensive efforts may be required. In any case, grout curtains are best for fractured rock. Grout curtains are expensive because they can generally only be injected into rocks made up of particles sand-size or larger. The details of emplacement as well as the advantages and disadvantages of each approach are examined in greater detail in (Smith et al., 1995). Horizontal barriers apparently remain a developing technology.

None of the technologies cited above are inexpensive, except, possibly for the installation of a rudimentary clay cap. An asphalt-hardened cap goes for 8–11\$/yd^2 (Smith et al., 1995) (All figures are in 1992 dollars). Soil-bentonite slurry walls cost 3–7\$/ft^2 at 0–30 feet deep; 6–11\$/ft^2 from 30–50 feet deep; and 9–15\$/ft^2 from 50 to 125 feet deep. Sheet piles cost between 16 and 28\$/ft^2. Depending on the method of emplacement the outlay for slurry walls can cost a great deal more than all of these numbers. Constructed wetlands have been used at a number of Superfund sites to immobilize metals as well.

Chemical treatments designed to decrease the solubility and mobility of metals include chemical oxidation of the metal, chemical reduction, and chemical neutralization. As noted earlier, many metals are most soluble and mobile in sub-neutral solutions. An obvious target for chemical oxidation is arsenic which is least soluble in its pentavalent state. Chemical reduction is most commonly targeted at CrO_4^{2-} which, as noted earlier, is quite insoluble once reduced to $Cr(OH)_3$. Conversion of chromate to insoluble $Cr(OH)_3$ is engineered with the addition of an electron donor such as H_2S gas or bisulfite salts. Chemical oxidation/reduction of radionuclides by bacteria has been demonstrated by Francis and co-workers (Francis, 1994). There is always a concern that geochemical engineering of contaminant redox state or pH will end up unbalancing some other aspect of the groundwater equilibrium and lead to unexpected results. For example, the engineered reduction of one hazardous substance might cause the reduction of another comingled contaminant that is more mobile in its reduced state. At the same time sorbed species might be desorbed due to changes in solution pH and/or redox state.

Two of the more routine tests used to demonstrate stabilization of wastes are the Toxicity Characteristic Leaching Procedure (TCLP), which is described in the Federal Register, and the California Waste Extraction Test (WET), which is described in the California Code of Regulations. The TCLP is meant to simulate the leaching which occurs in a landfill. For metal-containing solid waste, the solids (if not already paste-like) are ground to a particle size of less than 9.5 mm. An acetate buffer of approximately pH 5 in a quantity of 20 times the weight of the solid is used to extract the contaminant for 18 hours. The extract is then filtered through a 0.6 to 0.8 micron glass fiber filter, and analyzed for the contaminant of concern. The WET

Case Studies 87

test uses a citrate buffer of pH 5, a fluid to solid mass ratio of 10 to 1, requires particle sizes of 2 microns, and the leach lasts 48 hours. Deionized water may be used as the leaching solution if the wastes themselves are not capable of generating acids. The measured contaminant levels are then used in a simple transport model which takes some account of attenuation along a flow path to guage whether or not a waste poses a health risk to receptors.

The outright failure and/or great expense of existing engineered approaches and the uncertain prospects for new "great leaps forward" are adequately captured by the Natural Research Council's (1994b) recommendation that EPA invest more time and effort in informing the public about poor prospects. Since engineered efforts often have little effect on contaminant levels at depth it is useful to examine whether soils can erase toxicity on their own. The rest of this chapter will outline cases where natural attenuation of contaminants in soils and groundwaters has been demonstrated in the field.

PETROLEUM HYDROCARBONS

Perhaps the best-documented example of petroleum hydrocarbon degradation comes from a USGS study of a site in northern-central Minnesota where crude oil was accidentally released from a pipeline. A portion of the oil was sprayed over a large area while another portion accumulated as free product (see Figure 7.2). Subsequent biodegradation of the oil has been examined to understand how a groundwater reacts in a macroscopic sense to the perturbation of its chemical equilibrium. This example is important because it illustrates how the secondary effects of hydrocarbon degradation (e.g., shifts in bicarbonate and pH levels) can provide useful fingerprints of the primary process of intrinsic bioremediation. Figure 7.3 illustrates the geochemical zonation at depth. Zone I is upgradient from both the spray zone and the oil body and is pristine groundwater, which is a dilute $Ca-Mg-HCO_3^-$ water. The composition of the latter is determined largely by equilibrium with calcium carbonate minerals (Baedecker et al., 1988). In zone II aerobic degradation of spray oil has raised carbon dioxide levels, resulting in a lower pH and calcium carbonate dissolution, as evidenced by higher levels of Ca and bicarbonate in the sampled waters. The petroleum hydrocarbons detected in zone II are highly degraded.

Zone III, the section in contact with the separated phase, is highly anaerobic. No nitrate or sulfate can be detected in the groundwater. Instead, high Mn^{2+} and Fe^{2+}, as well as methane are found. Stable carbon isotope analyses (see above) also point to methane fermentation. High Si levels in the water are thought to arise from accelerated dissolution of silicate minerals by organic acids which are breakdown products from the petroleum (Bennett et al., 1993).

Zone IV is downgradient where iron and Si decrease back to background levels due to oxidation of Fe^{2+}, and oxidation of the organic acids which carried the Si in the first place. Figure 7.4 shows the concentration of benzene, ethylbenzene, and toluene as a function of distance downgradient, and relative to a non-reacting, conservative tracer. Obviously, there is a sink for each of the hydrocarbons. Given a choice between sorption and biodegradation, the latter appears most reasonable.

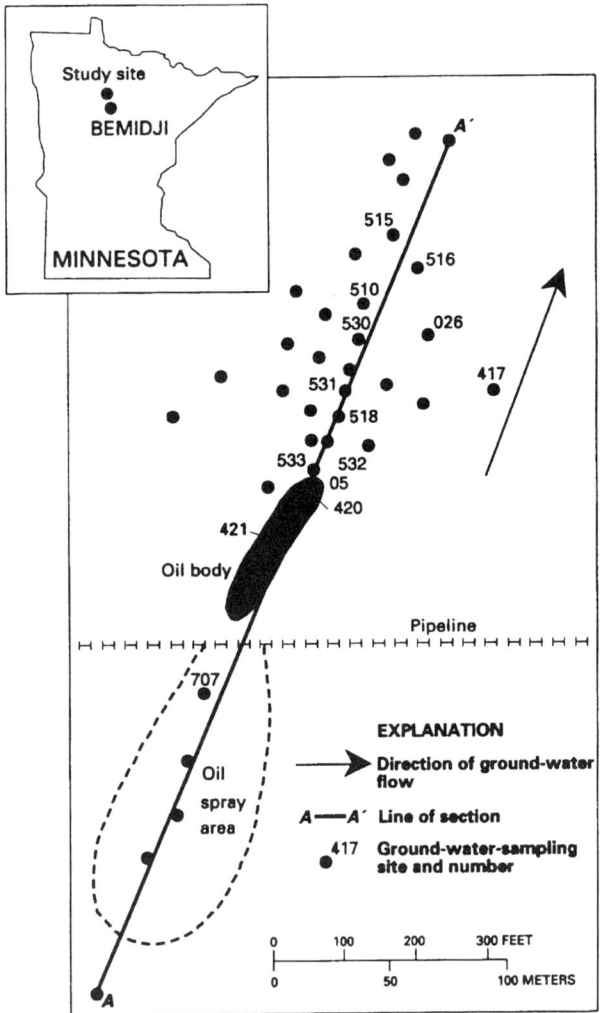

FIGURE 7.2 Map of USGS Bemidji site. (From Eganhouse et al., 1994).

Ethylbenzene sorbs more to aquifer organic matter than toluene, which sorbs more effectively than benzene. Because the sink affects the chemicals in different order, i.e., toluene more strongly than ethylbenzene, which is in turn removed more completely than benzene, sorption is probably not the primary sink. Consequently, biodegradation is thought to account for the curves.

Degradation of a contaminant molecule can be masked or mimicked by natural processes (Siegel et al., 1992) (for example degradation of naturally occurring organic matter might give similar trends to the example given above). It is therefore important to have a "control", or, equivalently, a non-contaminated sampling wherein the base-level biogeochemical reactions can be calibrated.

Case Studies

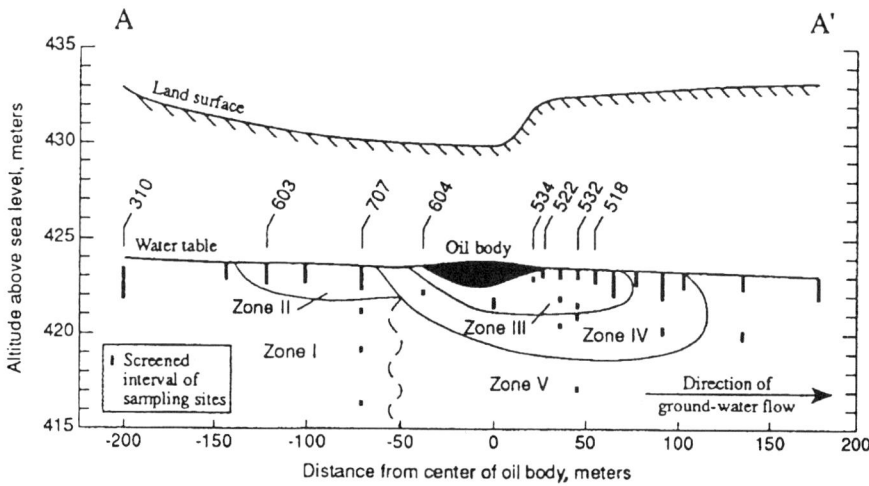

FIGURE 7.3 Geochemical zonation at USGS Bemidji site. (From Eganhouse et al., 1994).

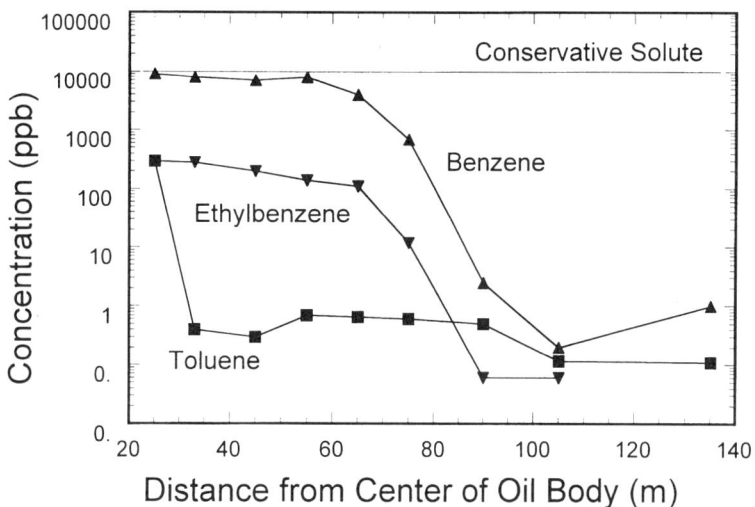

FIGURE 7.4 Benzene, ethylbenzene, toluene, and conservative tracer concentrations as a function of distance from the center of the plume at the USGS Bemidji site. (From Eganhouse et al., 1994).

BTEX compounds are non-chlorinated, and therefore not particularly resistant to degradation. A wealth of evidence suggests that the BTEX compounds are very rapidly degraded by natural processes, and BTEX-contaminated waters made drinkable often within a few years. A revealing example comes from Dr. Henk Dop, an environmental scientist in the Netherlands:

About a year ago I started a "pump-and-treat" classical groundwater remediation at a site contaminated with jet-fuel. My consulting engineer had designed this wonderful system which included the re-infiltration of oxygen- and nutrient-enriched treated groundwater, this to stimulate oxic biodegradation. Sounded just fine. When the system started to operate, a problem developed within the treatment container, i.e., unexpected amounts of iron precipitating within the system. Indeed, it turned out that the Fe^{++} concentrations in the groundwater were rising most noticeably. Suspecting natural processes using Fe^{+++} as terminal electron acceptor, I ordered the oxygen enrichment/nutrient addition to be closed down, and also for the pump-and-treat system to be operated at 1/3 of the design specs — a purely intuitive choice. We also decided to altogether stop the pumps at a sub-site, this as a control. Since then, Fe^{++} is still on the rise, I've lost all BTEX-compounds, and total PHC load is also decreasing — more rapidly than the pump-and-treat effort can be held accountable for. The same for the control sub-site. Fe^{++} elevations check out nicely against this phenomenon. If the trend continues, I might meet regulatory demands within a few years — without any real effort. Sometime soon, I'll need to explain to the Admiralty (this being a Naval site) that I may have (in hindsight) been a bit premature at this site. It has a long history, and total spending (including the present contract) now stands at USD 1.7 million.... (Personal Communication — Dr. S. H. Dop, Netherlands Ministry of Defense).

One of the drawbacks to the interventionist approach mandated by CERCLA in the United States is that a baseline of background degradation is rarely established before backhoes, vapor extraction setups, and drill rigs are pointed at the contaminant. Instead, natural attenuation of the contaminants is generally quantified after the engineered efforts have failed.

Field evidence for rapid BTEX degradation is abundant (see Table 6.3). Laboratory confirmation of toluene degradation by denitrifying microflora include those of Dolfing et al. (1990), Evans (1991), Fries et al. (1994), Schocher et al. (1991). Toluene degradation by sulfate-reducing bacteria has been demonstrated by Rabus et al. (1993). Destruction of toluene by Fe-reducing bacteria has been shown by Lovley et al. 1989, and by fermentative bacteria by Vogel and Grbic-Galic (1986), Grbic-Galic and Vogel (1987), and Edwards and Grbic-Galic (1994).

Rice et al. (1995) surveyed the LUFTs in California and found that, out of 12,151 public water wells tested, in less than 0.5% (48) could benzene be detected. Out of all 28,051 LUFTs only 136 were found to have affected drinking water wells. Many of the latter were private domestic wells close to the source. The reason that LUFT plumes have contaminated such a tiny portion of California drinking water is that soil microorganisms severely limit the movement of plumes by digesting the balance of the contaminant. Rice et al. (1995) cite data from a number of counties in California showing that contaminant plumes rarely exceed 250 feet and stabilize (or even begin to retreat) quickly. Once contaminant sources were removed, and plumes stabilized, natural degradation proceeded at substantial rates, sometimes as much as 50–60% per year. "Doing nothing" after the source was removed generally resulted in a ten-fold reduction in plume contaminant mass within 1 to 3 years (Rice et al., 1995).

In addition to endorsing the application of a risk-based corrective action (RBCA — pronounced "Rebeccah" — see below) approach to cleanup levels, Rice et al. (1995) recommended the following be done:

1. Minimize actively engineered LUFT remediation processes.
2. Once passive bioremediation is demonstrated and, unless there is a compelling reason otherwise, close cases after source removal to the point of residual fuel hydrocarbon saturation.
3. In general, do not use the UST Cleanup Fund to implement pump-and-treat remediation unless its effectiveness can be demonstrated.
4. Support passive bioremediation with a monitoring program.

MTBE is an additive used to decrease hydrocarbon and carbon monoxide emissions. It began to show up in gasolines in the late 1970s and appears to be an atypically recalcitrant component of gasoline. At present MTBE is tentatively classified as a possible human carcinogen. A recent study by the USGS (Delzer et al., 1996) documented the low-level occurrence of MTBE in groundwaters and stormwater samples in a number of cities and metropolitan areas.

Recall from Chapter 2 that petroleum production wastes are made up of a wide variety of naturally occurring organic molecules, often in a matrix of rock fragments. Much finds its way into soils, through spillage, uncontrolled dumping, or landfarming. There is an extensive literature attesting to the efficiency of indigenous organisms at degrading these wastes. Biodegradation rates observed in landfarming experiments are on the order of 130 to 600 g/m^2 per month (Bossert and Bartha, 1984). The degradation of petroleum hydrocarbons is covered in other sections of this book, and a number of more extensive coverages exist (e.g., Bossert and Bartha, 1984; Leahy and Colewell, 1990; Morgan and Watkinson, 1989; Oudot et al., 1989; Raymond et al., 1976).

LANDFILLS

As municipal solid waste (MSW) accounts for a large volume of the total waste produced, it is useful to examine the role of natural attenuation in its breakdown. At the same time, the filling up of MSW disposal sites is often cited as an "environmental crisis" precisely because much of our garbage doesn't naturally attenuate rapidly enough, for public satisfaction. In particular, the resistance of plastics to breakdown is often pointed out by commentators just before they state that " we are literally burying ourselves in garbage". Modern landfills are not designed to rely on natural attenuation. Early landfills did, by default. Nowadays, MSW is compacted on site, and routinely covered with dirt layers. The net effect is that degradation of organic matter rapidly becomes anaerobic (and slower), as the access of oxygen is limited. The movement of water is also limited, and the total volume of leachate produced by the slow percolation of rainfall through the pile is engineered to be small. Underneath modern landfills an impermeable clay liner or plastic membrane is emplaced to prevent the movement of leachate into groundwater. Instead the

TABLE 7.1
Products in MSW Landfills Ranked by Weight — 1988

Rank	Category	Million tons	Share (%)
1	Construction and demolition debris	61.0	20.9
2	Landfill lining and covering materials	58.0	20.0
3	Yard waste	31.1	20.0
4	Sewage sludge	16.0	5.5
5	Food waste	13.2	4.5
6	Paper shipping boxes	12.6	4.3
7	Miscellaneous durable products	10.5	3.6
8	Newspapers	8.9	3.1
9	Furniture and furnishings	7.5	2.6
10	Office paper	5.7	2.0
11	Other non-packaging paper	5.2	1.8
12	Books and magazines	4.6	1.6
13	Other miscellaneous non-durable products	4.6	1.6
14	Glass; beer and soft-drink containers	4.3	1.5
15	Folding paperboard cartons	4.1	1.4
16	Clothing and footwear	3.9	1.3
17	Other	39.8	13.7
	Total	291.0	100.1

From J. H. Alexander, *In Defense of Garbage*. Copyright 1993 by Praeger Publishers. With permission of Greenwood Publishers Group, Inc., Westport, CT.

leachate is collected and treated similarly to municipal sewage. This contrasts with the "open-pit" landfills of the past, where MSW was often tossed haphazardly into a low spot, where it was open to the elements, the latter in this case being wind, rats, birds, skunks, and the like. Because of the lack of covering the older landfills often remained aerobic. Natural degradation of organic matter (for example, cellulose from paper) was much more rapid, and, as a result, the landfills didn't fill up as fast (Alexander, 1993).

MSW in the United States can be roughly characterized by volume as (Alexander, 1993): 22.2% durable goods (appliances, tires, furniture, etc.); 34.0% non-durable goods (e.g., paper plates, diapers, shoes, ball point pens, etc.); 29.6% containers and packaging; 14.2% food, yard, and miscellaneous organic wastes (U.S. Environmental Protection Agency, 1990). Various classification schemes are used to pigeonhole the various waste forms (for example EPA doesn't include construction debris in their accounting). An arguably more specific characterization of MSW contents is that of Alexander (1993):

A couple of points drop out of Table 7.1: (1) the vast majority of MSW is natural (rock, wood, dirt, lawn clippings, sewage sludge and food waste account for 62% (Alexander, 1993)), and (2) paper in one form or another makes up much of the rest (~14.2%).

Case Studies

Unlike the relatively inert component (rocks, dirt, glass), the organic fraction can potentially biodegrade, and many assume that landfills are designed to aid the process. This is not true. The routine dirt coverings are designed to slow biodegradation so that leachate and gas (methane) production is limited (Alexander, 1993). It can, therefore, take decades for organic matter to decay in a landfill, that would be completely biodegraded in months if disposed under reasonably aerobic conditions. The net result is that leachate and gas production is stretched out over much longer periods of time.

Landfills wherein natural attenuation of the leachate was actively engineered have been proposed over the years (Bagchi, 1987). The basic ideal was to allow the leachate to seep through the unsaturated zone, where aerobic conditions would cause rapid breakdown of residual organic compounds (see Figure 7.5). Once the remaining leachate reached the water table and formed a plume in the saturated zone, organic degradation was assumed to continue. At the same time, it was estimated that adsorption of leachate metals, as well as dilution and dispersion, would then remove the remainder of the toxicity.

If leachates are cleaned by natural attenuation beneath the landfill, the generation of leachate inside the landfill becomes much less of a problem. In fact more rapid and more complete degradation of MSW might be achieved if the dirt coverings weren't emplaced and semi-aerobic conditions were encouraged. This would only be possible if the natural attenuation of leachate at depth prevented contamination of the adjacent groundwater. In fact, there is substantial evidence that natural attenuation of MSW leachates in some cases does just that.

Two of the earlier examinations of the chemical evolution of landfill leachate plumes were those of Reinhard et al. (1984) and Barker et al. (1986). who examined leachate plumes beneath two sanitary landfills near North Bay and Waterloo, Ontario. Both landfills sit atop permeable, sandy deposits. Both landfills had operated since the 1960s and had received primarily municipal waste, with a small fraction of industrial waste as well. The primary organic compounds which made it into the leachates of the two landfills came from decomposing plant material, and were made up of aliphatic and aromatic acids, phenol, and terpene compounds. In addition, compounds of industrial origin found in the plume included chlorinated and non-chlorinated hydrocarbons, nitrogen-containing compounds.

Breakdown under methanogenic conditions appeared to be the primary removal mechanism for dissolved organic carbon (DOC). Because of the reducing conditions, a number of aromatic contaminants were stable and persisted in the plumes hundreds of meters downgradient. At the same time a non-aqueous phase liquid (NAPL) was proposed as the source. Reductive dehalogenation under methanogenic conditions was postulated as a possible removal mechanism for the chlorinated organics. TCE and 1,1,1-trichloroethane were observed to be limited by biotransformation reactions to the immediate vicinity of the landfill (Barker et al., 1986).

A recent, extensive examination of what happens to MSW leachate components when they come into contact with groundwater is that of Christensen and co-workers (Bjerg et al., 1995; Heron and Christensen, 1995; Rugge et al., 1995). Rugge et al. (1995) examined the distribution of leachate components downgradient from the

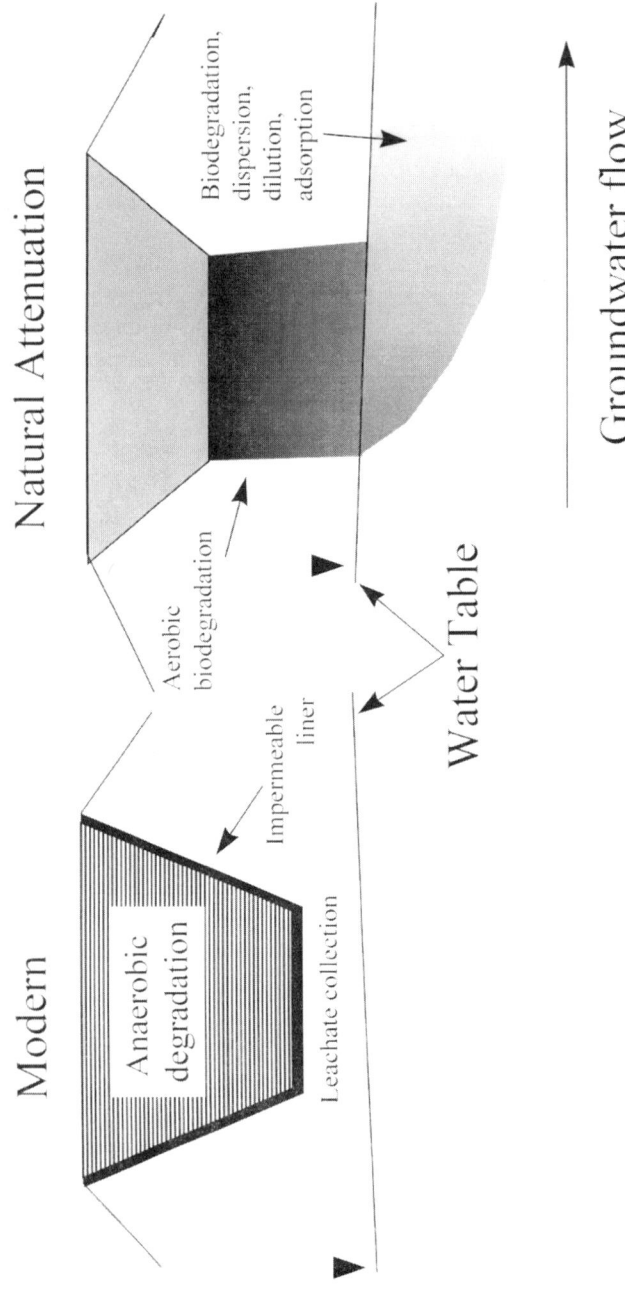

FIGURE 7.5 Schematic of MSW landfills; modern and natural attenuation. (After Bagchi, A., *Waste Management and Research*, 5, 453, 1987.)

Grindsted (Denmark) landfill. The landfill covers ~ 20 acres and is roughly 5 meters thick. Landfilling at Grindsted began in the 1930s and finished in the 1970s. Organic compounds detected at the source primarily included BTEX and napthalenes, though there was some evidence of pesticides as well. After accounting for dilution and minimal sorption, Rugge et al. (1995) documented that all contaminants were reduced below detection limits within 60 meters of the landfill.

The brief coverage above only points to a few of the cases where natural attenuation has been explicitly noted, and is by no means meant to be any sort of last word on landfill leachate chemistry. There are many sites where natural attenuation *hasn't* protected drinking water supplies. The effectiveness of natural attenuation primarily depends on what went into the landfill, the redox state of the landfill and underlying soils, and the distance the leachate had to travel to affect receptors.

METALS IN SOILS

To demonstrate natural attenuation of organic contaminants it must be shown that the activity of indigenous microorganisms is sufficiently robust to severely decrease contaminant levels before some point of compliance is reached. To demonstrate natural attenuation of metal contaminants it must be shown that metals, once sorbed to mineral surfaces, have since migrated into crystal lattices where they are effectively isolated from soil and groundwaters, and, consequently, the biota. A number of sources point to this sequestering of metals inside mineral surfaces. In laboratory experiments a large fraction of metals initially sorbed to mineral surfaces is often found to be very hard, or impossible, to remove without destroying the mineral itself.

Some important features of irreversible sorption of metals to mineral surfaces are outlined schematically in Figure 7.6. The vertical axis measures the fraction of metal, Me, sorbed to a mineral surface. The two vertical faces show: (a) what happens to sorbed metals the longer a metal-contaminated solution remains in contact with a mineral matrix, and (b) what happens to the sorbed fraction once it is confronted with fresh, metal-free recharge. The first case describes time-dependent sorption at a hot spot before the contaminant source has been removed. The latter case describes what happens after the hot spot and/or contaminant source is removed and fresh, metal-free solutions percolate through the soil. Note first of all that fresh recharge causes a net desorption of metals from soils at a rate described by one of the kinetic relations described earlier, in Chapter 5. In effect the surface tries to re-equilibrate with solution, releasing sorbed metals to achieve a thermodynamic equilibrium roughly described by a K_d. Generally less metal comes off than is predicted by a sorption K_d. Instead the K_d describing sorption increases over time. Because a significant fraction of the sorbed metal does not re-equilibrate with solution, this fraction is apparently no longer in contact with the solution, or alternatively, the sorbed metal has become a great deal more strongly sorbed. The mechanistic origins of this irreversibly sorbed fraction (henceforth termed Δ) will be explored in further detail below.

Generally, Δ becomes larger the longer metal-rich solutions are in contact with minerals. This is somewhat analogous to what happens when insoluble and/or strongly sorbed organics seep into matrix pores which are otherwise inaccessible to leach solutions. Although Figure 7.6 is schematic, it is useful as it shows that the

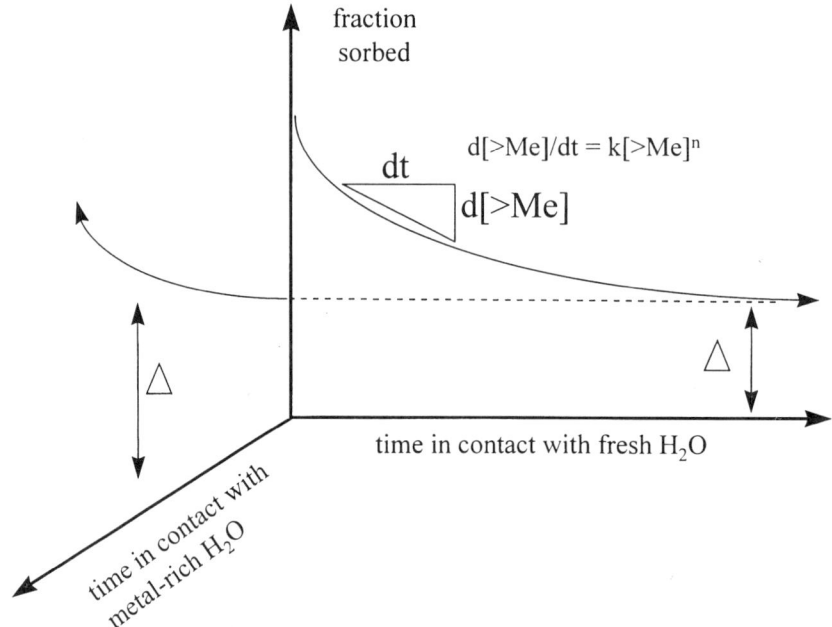

FIGURE 7.6 Irreversible sorption.

maximum amounts of contaminant removal that soil flushing might achieve; i.e., the non-Δ fraction is less than what is actually in the soil. To remove all of the contaminant the host mineral superlattice must be destroyed. In effect the Δ defines the most reasonable environmentally acceptable endpoint for cleanup. Soil treatment efforts (grout curtains, clay caps, reactive barriers, etc.) make little sense if their object (the Δ fraction) is already stabilized inside minerals.

Note though that the non-Δ, potentially removable, fraction is a maximum, as the actual yield will depend on the desorption rate. The curve in Figure 7.6 might become flat within hours to days (k is big), in which case desorption would tend to keep ahead of flushing efforts. On the other hand, the curve might not flatten for years to decades (k is small) in which case soil flushing efficiencies would be controlled by the desorption rate, and driven by the latter to very low levels.

Although irreversible adsorption of metals to mineral surfaces has been known for quite some time, the concept has not been applied to make cleanup decisions as widely as have the concepts of microbial degradation been applied to the cleanup of sites contaminated by organic contaminants. The object here is therefore to collect a number of the evidences for irreversible sorption of metals and to suggest how these might be applied in the future to metal-contaminated sites.

Cs SORPTION ONTO CLAYS

The case of Cs is one of the best studied examples of irreversible sorption (see e.g., Comans et al., 1991). Cs, although loosely bound to clay mineral surfaces at first,

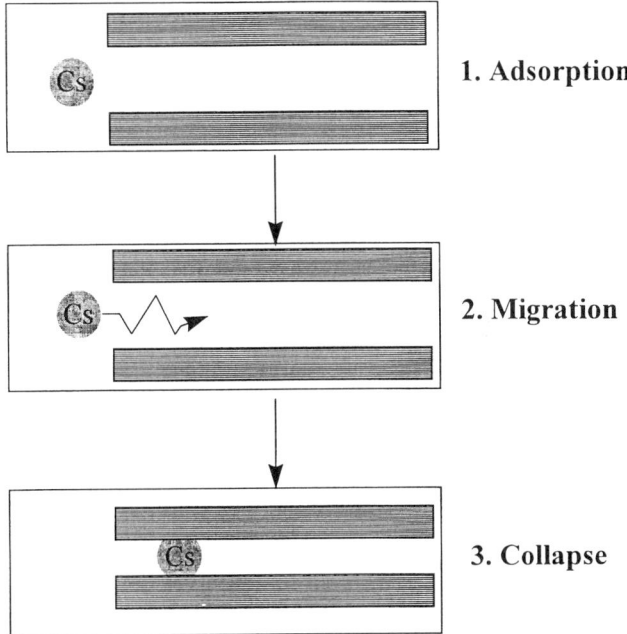

FIGURE 7.7 Sorption of Cs to clays.

given time tends to migrate to "high-energy", interlayer sites (see Figure 7.7)(e.g., Comans et al., 1991; Hsu and Chang, 1994; Khan et al., 1994; Ohnuki, 1994). Upon arrival at the latter the Cs loses any waters of hydration it carried with it and then orients itself in the site. For many 2-layer clays, this dehydration and site occupancy causes a local collapse of the clay structure. To subsequently remove the Cs requires either a re-opening of the layers, or destruction, of the mineral lattice. In effect, the move to high energy sites means an increase in the K_d. The K_d, to a greater or lesser extent, reflects the free energy difference between a sorbed molecule and the same molecule in solution. The movement of the Cs from the edges to the interlayer sites causes a net reduction in the free energy of the sorbed Cs, which translates to an increase in the K_d with time.

METAL SORPTION TO CARBONATE MINERALS

There is considerable evidence that adsorption of metals to limestone surfaces often results in their incorporation into crystal lattices (Davis et al., 1987). Zachara et al. (1991) allowed calcite surfaces to equilibrate with metal-rich solutions for 24 hours, and then equilibrated them with fresh solutions. Figure 7.8 shows the fraction of metals sorbed on calcite that were removable in an 8-hour leach period. Note that for the metals Cd and Mn less than 30% of the initial dopants were available for desorption. For other metals, recovery was somewhat more complete, but in no case was 100% recovery achieved. Desorption efficiencies are plotted against the negative free energy of hydration of the various metal cations, and the correlation between

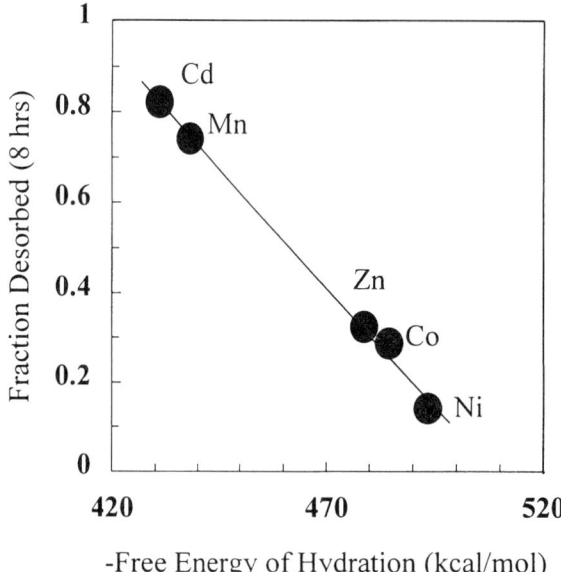

FIGURE 7.8 Metal desorption from calcite. (From Zachara, J. M. et al., *Geochim. Cosmochim. Acta,* 55, 1549, 1991. With permission from Elsevier Science Ltd., Kidlington, U.K.)

the two is thought to point to the mechanism of metal incorporation at the calcite surface. Namely, hydrated cations, once sorbed, are thought to be incorporated at Ca sites and subsequently overcoated as the mineral recrystallizes (see Figure 7.9). Dehydration at the mineral surface is thought to be a critical step in the sorption process as it decreases the effective radius of the cation, making it easier for metal substitution for Ca sites at the mineral surface to occur. Fuller and Davis (1987) observed a similar slow incorporation of Cd^{2+} onto a calcareous aquifer sand and noted that the amount sorbed increased slowly the longer the two were in contact.

URANIUM SORPTION TO SOILS

Figure 7.10, showing U adsorption and desorption measurements of Barney (1984), dramatically illustrates the often observed non-reversibility of adsorption. Specifically, K_d's measured from the adsorption of uranium onto interbed materials of the Columbia River basalts are quite different (and generally much lower) than K_d's measured through desorption of uranium from the same materials. Desorption at 60°C is shown under reducing and oxidizing conditions. The diagonal lines having slopes of 1 represent K_d's measured separately in sorption experiments done from metal-rich conditions. If sorption of uranyl was reversible then the desorption data shown in Figure 7.10 should fall on the diagonal lines. Under oxygen-poor (reducing) conditions effectively no desorption of sorbed uranium, neptunium, technetium, or selenium was detected. Under oxidizing conditions neptunium adsorption was still somewhat irreversible, whereas uranium adsorption under the same conditions

Case Studies

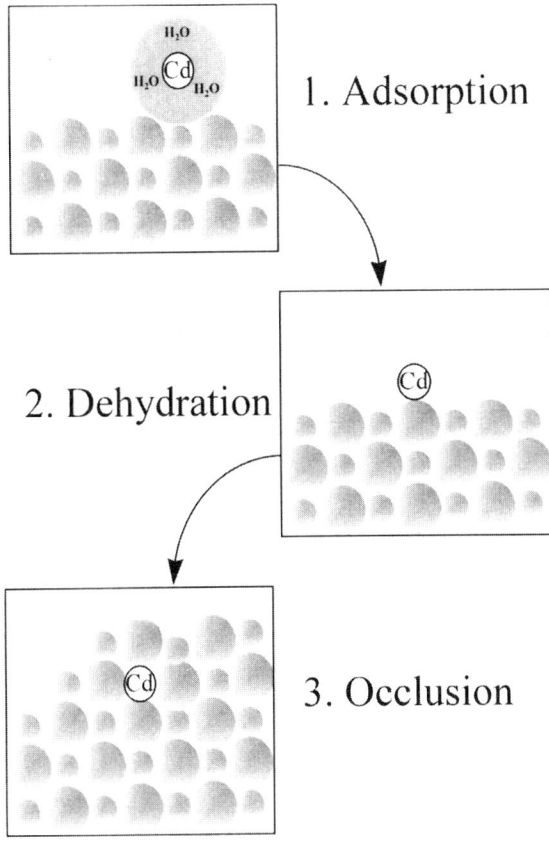

FIGURE 7.9 Overcoating of Cd on calcite.

was highly irreversible. Under oxidizing conditions desorption was scant, and nowhere near that predicted by the equilibrium K_d. The only way to achieve a reasonable fit between the data and theory would be to allow the K_d to increase over time. In a mechanistic sense what this means is that sorbed uranium migrated to sites which were a great deal stronger than those initially occupied. Under reducing conditions the mismatch between adsorption and desorption is even worse. Effectively no desorption was observed, and the K_d measured from the desorption side is much higher than that measured from the adsorption side. In effect sorption resulted in a net reduction of contaminant mass available to solution. All transport codes which model sorption use fixed K_d's, which assume reversible sorption and no irreversible uptake of metals into mineral surfaces. For many metals this has the effect of exaggerating the calculated rate of contaminant transport in the subsurface and overestimating the actual amount of bioavailable contaminant mass. The net result of a "reversible K_d, no metal uptake" approach is to exaggerate the amount of cleanup required and its ease of achievability.

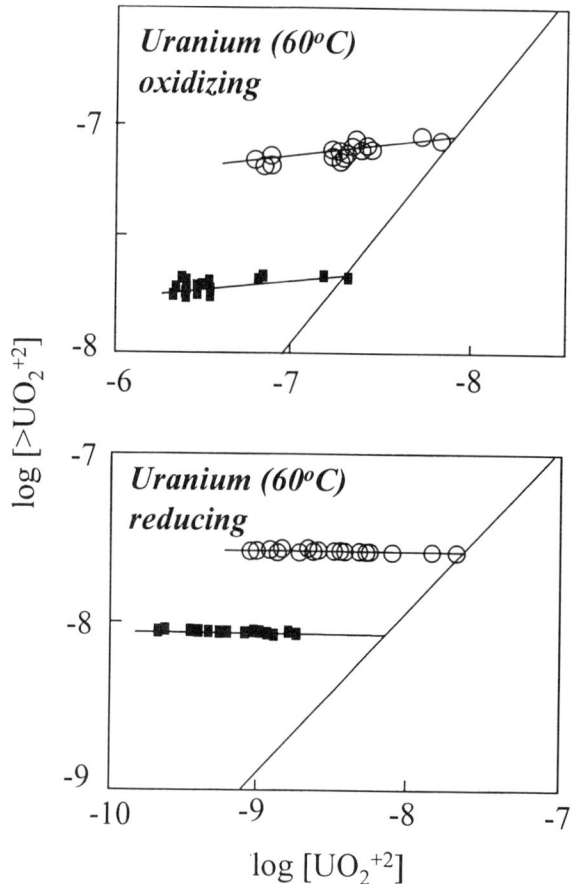

FIGURE 7.10 Desorption of U from Columbia River Interbed Materials. (From Barney, G.S., *Geochemical Behavior of Radioactive Waste*, p. 3, 1984. Copyright 1984, American Chemical Society. With permission.) The vertical axis is the amount of uranium sorbed (mol/g). The horizontal axis is the dissolved U level (mol/l).

METAL SORPTION ON IRON (HYDR)OXIDES

Iron hydroxides tend to recrystallize in near-surface oxidizing environments. In the process their surfaces adsorb many metals, and the recrystallization process has the effect of removing metals from solution semi-permanently. If a system-wide change in redox state or pH were to occur and increase the solubility of the mineral, and cause it to dissolve, the sequestering would cease, and the once-bound metal might make it back into solution. A good example of this is illustrated by Figure 7.11 which shows a back-scattered electron image of Pb and Fe sequestered in the pitted surface of a basalt from an exposure in Tempe, Arizona. The lead is probably from automobile exhaust. The iron came from the surrounding soil or from oxidative

Case Studies

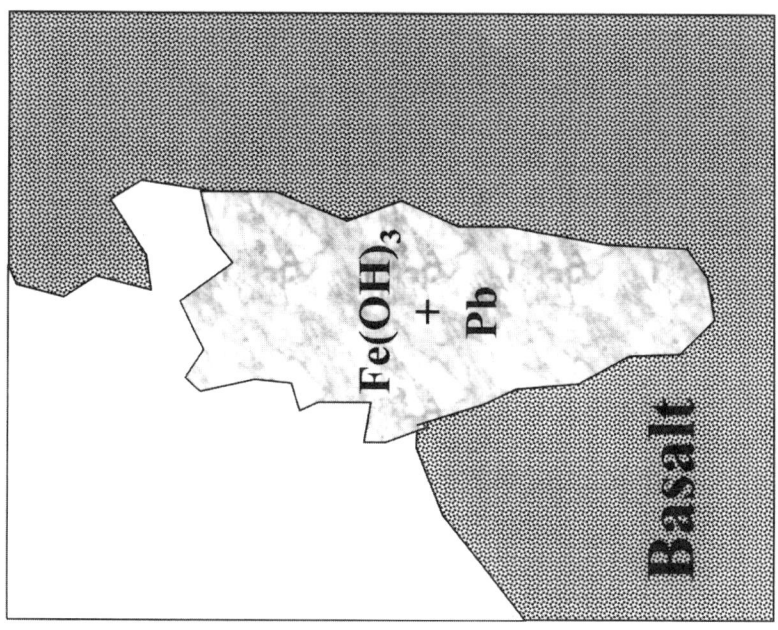
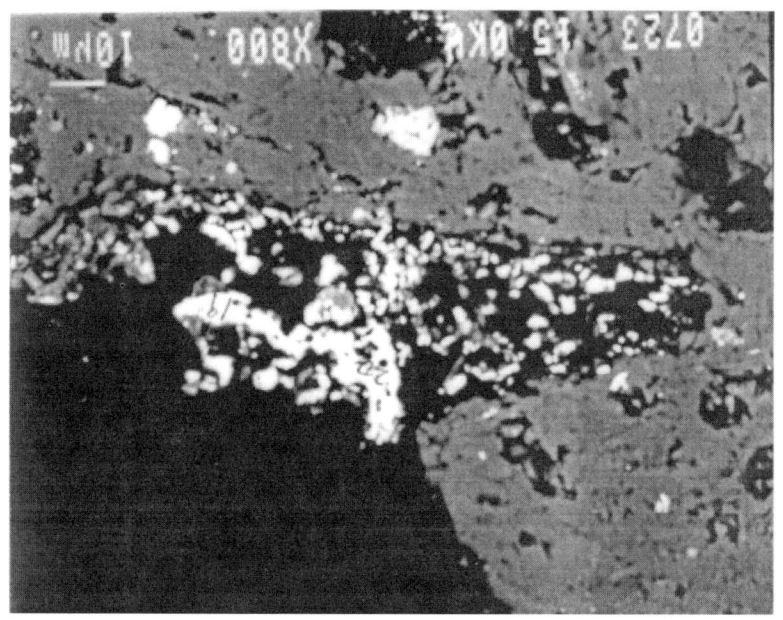

FIGURE 7.11 Lead occluded by Fe-hydroxide in crack of basalt. (Courtesy of R. I. Dorn — Arizona State University).

dissolution of the divalent iron-containing minerals in the basalt. When synthetic soil solutions are passed over the basalt, only trace levels (~3 parts per billion) of Pb show up in the leach solution. When the iron hydroxide coating is selectively leached away with a dithionate-citrate solution beforehand, and the basalt subsequently leached again with the synthetic soil solution, ppb levels of lead are detected. On the other hand, the dithionate-citrate leach is observed to have lead levels three orders of magnitude greater (~10 parts per million), which confirms the observation that there is lead contamination at the mineral surface. It's unleachability suggests that it is occluded inside the iron hydroxide lattice (there may be some carbonate mineral inclusions as well). Obviously, for this contaminant-soil pair the Δ is very large.

LEAD IN SOILS

The controls on irreversible sorption outlined above are useful in examining field evidence which indicates that many metals never make it back into soil and groundwaters once added by anthropogenic processes (human activities). Lead is a particularly useful example of this. In 1970, 253,000 metric tons of lead were produced (U.S. Bureau of Mines, 1972), and a significant amount made it into gas tanks, and, subsequently, into the atmosphere. When it came down, most of it landed on soils, through which aquifers recharged and streams drained. Very little of this massive flux of lead actually shows up in groundwater or streams. Recent high-precision measurements of Wang et al. (1995) point to near-complete retention of lead by a northern (U.S.) hardwood forest soil. This is not surprising from the discussion in Chapter 2, which pointed to the tendency of lead to sorb (or form insoluble hydroxycarbonates). In general, metals which form insoluble minerals also sorb strongly. Another important aspect of the retention of lead is that it appears to be irreversible to a degree. The use of leaded gasoline has steadily declined in the U.S. over the past 20 years, and the flux of lead to soils has likewise decreased. Although progressively fresher (more lead-free) rainfall is making it into soils there appears to be little freeing up of sorbed lead to solution, as might be predicted by equilibrium adsorption/desorption or mineral solubility controls. The rate of lead release is so slow that desorption appears to be negligible. The same behavior is also observed for many radionuclides, as will be seen in the following section. It should be pointed out though that when erosion or construction sends small silt-sized particles and colloids into streams, they carry the sorbed lead with them. Lead pipe solder represents another potential contributor of lead to drinking water as it corrodes into drinking water when the pH (and/or the bicarbonate level) of the latter is low. In theory, it is possible to control pH and bicarbonate levels to minimize lead liberation from pipes (Schock, 1980). One feature of insoluble metals deposited from above (lead and plutonium — see below) is that, since they don't move rapidly into the soil profile, their primary threat is often through ingestion of particulate matter. In this way they can bypass any negative bioaccumulation (see below), and consequently pose a potentially greater health risk. The nice aspect is that if ingestion is a potentially important health threat, remediation can often be done with a backhoe, followed by isolation and capping elsewhere.

TABLE 7.2
Retardation Factors for Important Radionuclides in Potential Repository Rocks

Element	Granite	Basalt	Tuff (volcanic ash)	Shale or clay	Salt
Sr	20–4000	50–3000	100–100000	100–100000	10–50
Cs	200–100000	200–100000	500–100000	200–100000	40–100
I	1	1	1	1	1
Tc	1–40	1–100	1–100	1–40	1–10
U	20–500	50–500	10–400	50–2000	20–100
Np	10–500	10–200	10–200	40–1000	10–200
Pu	20–2000	20–10000	50–5000	50–100000	40–4000
Am	500–10000	100–1000	100–1000	500–100000	200–2000
Ra	50–500	50–5000	100–1000	100–200	20–50
Pb	20–50	20–100	20–100	20–100	1–20

From Krauskopf, K. B. et al., *Radioactive Waste Disposal and Geology,* Chapman & Hall, New York, 1988. With permission.

RADIOACTIVE SOILS

It would be nice if all of the radioactive contaminants in soils worldwide could be removed. Nevertheless, a vast wealth of evidence suggests that the effort is probably unnecessary, even if it were doable, and it almost certainly isn't. The testing of nuclear weapons in the atmosphere 30 to 40 years ago put a sufficiently large quantity of long-lived radioactive elements into the atmosphere, and subsequently into soils, that scientists have been able to track and work out the details of their fate and transport. Consequently, it's not hard to predict where radionuclides in soils and groundwaters are going. For most of them, the answer is "not far". Table 7.2 gives a compilation of retardation factors, R_f, for important radionuclides in a number of rock types that have been considered at one time or another as a site for nuclear waste. The numbers come from a variety of sources and are useful because they point up some general trends, the most important being that, with the exception of iodine and technetium, in most soils there is at least one common constituent to which radionuclides will stick. This means that they don't move all that far in the environment, which is fortunate from a public health point of view. At the same time, it means that cleaning soils of radionuclides is going to be a near-impossible task. Recall from the hypothetical case in Figure 6.2 that a relatively low contaminant K_d of 0.3 mL/gm can double cleanup costs. K_d's of radioactive contaminants in soils are routinely orders of magnitude higher still.

Approximately 100,000 curies of plutonium were released into the atmosphere by about 400 nuclear explosive tests conducted by the U.S., Great Britain, the U.S.S.R., China, India, and France since 1945. This amounts to roughly 6,000 kilograms of plutonium, spread out over soils and natural waters (Facer, 1980). A number of accidents and intentional releases have also added plutonium to the environment. Plutonium came down from the skies as plutonium oxide, which is quite insoluble. It has moved through soils at somewhere between 0.1 and 1.0 cm per year (this number may be high as a number of evidences (see e.g., Nuclear Energy Agency, 1981) point to even slower transport). The plutonium dispersed from the Trinity explosion, the first nuclear detonation, done east of San Antonio, New Mexico on July 16, 1945, moved between 2.5 and 7.5 cm down into the soil column in the subsequent 27 years. A lot of this may be due to surface reworking.

Plants have picked up fallout plutonium primarily through roots (some landed as dryfall and was sequestered on leaves). This is not to imply bioaccumulation, though, as plants and animals don't take up plutonium readily. The level of plutonium in a plant in contaminated soil is generally one ten thousandth or a millionth of the plutonium level in the soil. By the same token a grazing animal which consumes that same plant will retain approximately one ten thousandth of the amount of plutonium the plant has, and a human eating the animal will in turn have roughly one ten thousandth of the plutonium that the grazing animal does. This means that roughly a trillionth of the plutonium in soil will make it into a human eating meat, and a hundred millionth for someone eating crop plants. The cumulative discrimination of plants, animals, and humans against uptake of plutonium is extreme (see Nuclear Energy Agency, 1981). ^{241}Am is a decay product of uranium, and for all intents and purposes, its uptake looks to be about as unfavored as plutonium's.

A wide variety of radioactive isotopes were generated at the Hanford Reservation in the state of Washington during the production of nuclear bomb components. Many of the process streams containing low and intermediate-level radioactive waste ended up in settling ponds. Some were vented directly to the Columbia River. Of the radioisotopes which made it into soils, a number have moved little since then (Nuclear Energy Agency, 1981). Plutonium in soils contaminated by radioactive wastes produced at Los Alamos, New Mexico has been picked up by plants, though there appears to be as low as a one part in a thousand discrimination factor (the number is as high as one to one, though). Plutonium in the soils at the Rocky Flats plant outside of Denver, Colorado has moved a matter of centimeters through the soil column.

Laboratory confirmations of this immobilization of radionuclides at mineral surfaces is widespread. Payne et al. (1994) showed that uranium sorbed to ferrihydrite became irreversibly attached upon recrystallization of the latter to more stable goethite. Sims et al. (1996) found that ~10% of uranium initially sorbed to a sandstone in a core-flushing experiment was irreversibly sorbed, and postulated that the initially sorbed uranium was transformed to a less-leachable surface precipitate. In addition to the uranium and cesium results covered above, cobalt sorption to ferromanganese oxides is found to be, to an extent, irreversible (Park et al., 1992). Irreversible sorption of cobalt and nickel onto clay minerals was observed by Grütter et al. (1994) (Sr and Ba sorption was observed to be reversible).

In conclusion, many of the important radionuclides are so tightly bound to soil material (clays, (hydr)oxides, and organic matter) that they may not desorb over environmentally relevant time scales (time spans much less than their half-life). In effect, engineered immobilization, or soil washing, isn't needed as the contaminants are already immobilized (e.g., Bunzl et al., 1995). Even if plutonium and americium weren't sorbed strongly to soils, but instead were available to organisms, a great deal of evidence says that the plants and animals wouldn't take them up anyway.

Exceptions to the strong-sorbing radionuclide rule may be alkali cations (strontium and barium) and anions, such as iodide and pertechnetate (under oxidizing conditions), though there are a number of specific minerals capable of sorbing the former (Balsley et al., 1996). Note though that the presence of chelating ligands (e.g., EDTA) has the effect of decreasing sorption over at least short time scales.

A few caveats are in order before going further. Sometimes metals don't sorb in the field at the levels lab measurements predict. Sometimes they don't sorb irreversibly. Generally the first happens when a desorption-enhancing ligand is present. (Recall that those ligands which increase the solubilities of minerals often decrease the tendency of their metal components to sorb as well.) Means et al. (1978) showed that a number of radionuclides (in particular ^{60}Co) had migrated from low level disposal units at Oak Ridge National Laboratory in Tennessee much more rapidly than would be predicted by standard transport models due to the presence of EDTA. EDTA was a chelating agent used in the processing of the wastes which was buried with the waste, and subsequently envisioned to prevent the sorption of ^{60}Co. Unlike many of the other organic chelating agents which have relatively short half-lives in soils, EDTA is fairly recalcitrant and has a half-life of around 20 years. Other complexing agents might have the same effect on sorption and transport of other metals. However, note that the transport role of chelating agents will be limited by the fact that any sorption of the chelate-free metal to a mineral surface will cause a progressive fractionation of metals out of metal-chelate complexes and onto mineral surfaces. This has been shown for sorption of Cd-EDTA by Fuller and Davis (1987) and relies on the fact that an initial sorption of an unchelated metal to a mineral surface will decrease the amount of free metal in solution. The solution will re-equilibrate to restore the ratio of free metal to chelated metal, fixed by the metal-chelate equilibrium constant, by converting a portion of the metal-chelate complex to free chelate + free metal. The process of re-equilibration and metal-chelate breakdown begins again when the free metal sorbs. The net effect of such a process is to remove chelated metals from solution and onto surfaces. Conceptually, it requires that equilibration between the chelated metal and free chelate + metals be rapid, and that at least some fraction of the total amount of metal in solution be unchelated. If equilibration is rapid, the removal of metals to mineral surfaces should be most rapid for those which are least strongly chelated.

Drastic changes in soil solution composition might cause a decrease in metal binding to mineral surfaces. Oxygenated waters might free up uranium that was bound up under reducing conditions which prevailed earlier. Large-scale decreases in pH would favor desorption of many metals, as well as the dissolution of calcium carbonate or iron hydroxide hosts. Nevertheless, there are a number of processes covered earlier (gas dissolution/exsolution, mineral dissolution and growth, etc.) which tend to stabilize the gross chemical makeup of soils and groundwaters, and limit excursions from a steady-state.

URANIUM MILL TAILINGS

Uranium mill tailings in the U.S. are, for the most part, found in remote, arid areas. They can potentially contaminate groundwaters through leaching. Primary concerns are water-borne uranium and trace metals. Some sites still retain the residual components of the solution (e.g., sulfuric acid) used to originally leach and process the uranium ore. At the same time, the continual fluxing of radon to the atmosphere is a cancer hazard. Engineered caps are useful tools for delaying radon long enough that most of it decays to less mobile, and less dangerous, daughter products. There is ample evidence, though, that interaction of the fluids which leach out of the bottom and interact with adjacent groundwaters removes much of the environmental threat at many sites.

One of the best studied tailings piles is at Riverton, Wyoming. The tailings pile at Riverton operated from 1958 to 1963, resulting in the production of 90,000 metric tons of processed ore and spent leach solutions (White et al., 1984). The tailings pile occupied approximately 60 acres before it was removed, and it was 50 to 100% saturated due to a combination of residual pore water, recharge from rainfall, and intrusion of groundwater. Figure 7.12 shows profiles of dissolved chemicals through the tailings pile. Beneath the pile the rocks are an alternating series of sand and gravel-bearing alluvium and shales. Focusing for a moment on the variation in pH, note that the pH reached a minimum of around pH 1 at 2 to 3 meters depth, but rose to near background levels (pH 7.5 to 7.6) below 4 meters depth. The relatively low pH in the top two thirds of the pile was caused by the leftover sulfuric acid from the ore leaching. The neutralization of the acid seen in the bottom third came about due to reaction of the latter with calcium carbonate minerals in the underlying rock. The relevant reaction is:

$$2CaCO_3 + H_2SO_4 \Leftrightarrow 2Ca^{2+} + SO_4^{2-} + 2HCO_3^-$$

Confirming evidence of this scenario is the elevated calcium and bicarbonate levels seen at the bottom of the pile. The drop in aluminum and iron levels came about at depth due to the fact that aluminum and iron are least soluble at near neutral pH, and form their own hydroxide minerals. Note the drop in copper, selenium, and zinc levels, which paralleled the appearance of iron and aluminum hydroxides. Uranium levels were low at the bottom of the pile because of dilution by groundwater (White et al., 1984), and there is concern about the nature and extent of uranium migration from the site. Because uranium can't be biodegraded, and is relatively soluble (at least in its oxidized state) the only likely sinks for it are dispersion and adsorption. Bryan et al. (1996) used experimental measurements of uranium sorption onto aquifer material from Riverton as input to generate a hydrogeochemical model of uranium transport from beneath the tailings piles and examine the interplay between the two. Dispersion appears to be a less important control over transport, whereas the nature and extent of adsorption are critical. Measured K_d's were somewhat low (1.38 to 4.48 mL/g), but the surface area of the aquifer was quite high. The conclusion of Bryan et al. (1996) was that over the next century the vast majority of the uranium would become tightly bound to the soil in spite of the relatively low K_d. In other words, there would be no "smearing" and dilution of the plume away from the site. Instead the uranium is expected to remain sequestered on mineral surfaces.

Case Studies

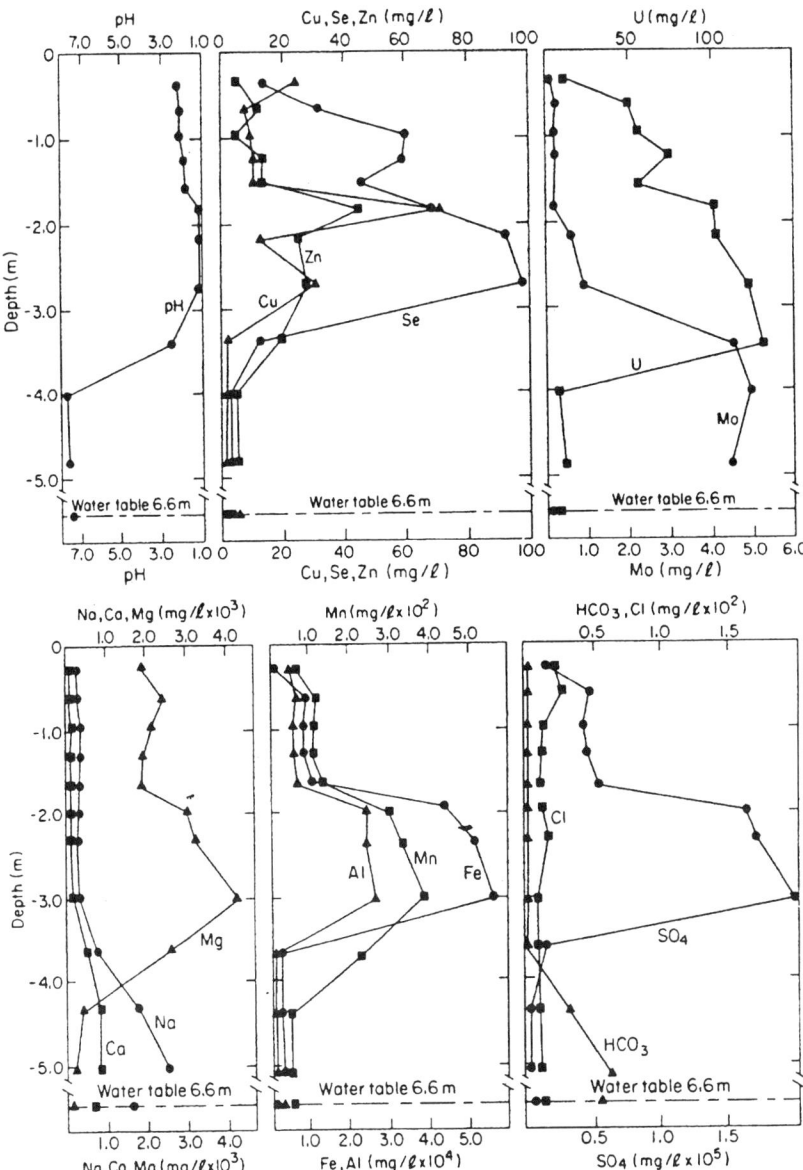

FIGURE 7.12 Vertical chemical distribution of pore waters in the tailings pile at Riverton, Wyoming. (From White, A. F. et al., *Water Resourc. Res.*, 20, 1743, 1984. Copyright American Geophysical Union. With permission.)

TABLE 7.3
Important Radionuclides in
LLRW and Their Half-Lives

Isotope	Half life (in years)
^{3}H	12.33
^{59}Ni	80 thousand
^{63}Ni	100
^{90}Sr	29
^{99}Tc	210 thousand
^{129}I	1.6 million
^{137}Cs	30.17
^{226}Ra	1600
^{232}Th	14 billion
^{234}U	240 thousand
^{238}U	4.5 billion
^{241}Am	432
^{241}Pu	14.7
^{14}C	5400

RADIOACTIVE WASTE

Natural attenuation is explicitly relied upon in the disposal of low and high-level radioactive waste for a couple of reasons. It is impossible to assure the performance of engineered structures (casks, liners, repository backfills) for the often long periods of time (hundreds to hundreds of thousands of years) required for the various radionuclides to decay. At the same time, the steady decay of radionuclides means that disposal needn't be "final", as isolation from the biosphere alone reduces potential health risk. The ten half-life rule is a useful means for defining how long is long enough. Current regulations call for at least 100 years of institutional control over LLRW disposal facilities (Gershey et al., 1990). In this time many of the short-lived radioisotopes will decay to insignificant levels. The primary ones which will remain are listed in Table 7.3.

The radioisotopes in Table 7.3 (plus the short-lived isotopes) represent more than 99.999% of the radioactivity in the inventories at the Barnwell, Beatty, and Richland LLRW disposal facilities.

LLRW disposal will by-and-large rely on burial in cement. Cement degrades over time to produce relatively high pH solutions, which makes many radionuclides insoluble and/or strongly sorbed. This sets some limits on the levels of the latter which will make it into soils once a LLRW facility degrades, some time after 100 years of institutional control. Standard geochemical calculations of radionuclide solubilities, combined with laboratory measurements of sorption to cementitious material and soil, suggest that the only radionuclides likely to make it into groundwaters, and pose a health risk, are ^{3}H, ^{99}Tc, ^{137}Cs, ^{129}I, and ^{90}Sr (Brady and Kozak,

1995). The remainder of the radionuclides either form insoluble minerals, sorb very strongly to cementitious materials and soils, or do both. After sorption (and mineral growth) the only sink remaining for these is dilution. As 100 to 1000 volumes of water are needed to degrade a cementitious repository (Berner, 1992) there is the possibility that dilution might reduce ^{90}Sr, ^{137}Cs, and ^{3}H to acceptable levels, which leaves ^{99}Tc and ^{129}I. Note though that these three constitute an extremely small collective dose which is calculated to be orders of magnitude less than the starting dose.

Because of the higher radioactivity and longer half-lives of the radionuclides, high level waste will probably be disposed in deep geologic formations. Presumably, the added distance between the latter and potential receptors would make the movement of radionuclides from one to the other very long. There are a number of exhaustive examinations of geologic disposal of radioactive waste (Brookins, 1984; Krauskopf et al., 1988), so only the broad outlines of the problem are covered here. It takes somewhere between 10,000 and 100,000 years for the radioactivity of fuel elements from nuclear reactors to decay to the level of a uranium ore body. The latter is generally taken as the target for HLRW disposal efforts.

Fuel rods are removed from nuclear reactors after some fraction of the ^{235}U originally in the rods has fissioned (split) to produce heat. What remains in the spent fuel rods is residual uranium, as well as a number of fission and activation product isotopes. In some cases the fuel rods are reprocessed to extract and reuse the fissionable material. The early activity of waste from reprocessed fuel is principally fission products having relatively short half-lives. Consequently, reprocessed wastes lose the vast bulk of their radiotoxicity within a thousand years. Unreprocessed fuel takes close to a hundred times longer to decay as far.

There are a number of reasons why the deep Earth was picked as the place for this long wait (see e.g., Faure, 1991), in addition to the extended travel time angle noted above:

1. Rocks can dissipate the high heat output which will continue to emanate from the waste for many years,
2. Minimal maintenance is required over the long haul,
3. Terrorist attacks and thefts are harder to pull off in deep holes underground.

There are potential hazards involved, including earthquakes and/or volcanic eruptions, both of which might breach a repository and release the waste from its canister. Groundwaters might also corrode the waste-containing canisters and move the waste into the environment prematurely. Human intrusion scenarios also include the possibility of the waste being dispersed by the drilling activities of some future civilization (Presumably, any civilization able to poke holes thousands of feet under the ground should also be able to detect, and subsequently avoid, radiation as well). Note though that most of these events (plus a few others — for example, dispersal by hurricanes, terrorists, glaciers) are just as likely happen if the waste is left near the surface, where it is now. There have been recent cases of scavengers breaking into above-ground storage areas and dispersing radioactive isotopes.

In general, underground disposal locks in a relatively long travel time to the surface, and the biosphere. The uniformly high K_d's listed in Table 7.2 testify to the very slow and difficult journey an escaped radionuclide will have in front of it if it is forced to move through rocks for any great distance. The sluggish movement of fallout material in soils confirms this trend (see above). There is even more compelling evidence pointing to radionuclides as the slowpokes of environmental transport, and this comes from the "Natural Reactor" at Oklo, Gabon.

Roughly 2 billion years ago, a uranium-rich rock there went critical, that is, a chain reaction of fissioning ^{235}U caused the rock body to heat up, whereupon water began to circulate through to moderate the reaction, directly analogous to a modern-day nuclear reactor. Unlike the newer ones which are relatively short-term affairs, the Oklo reactor stayed in business for roughly 500,000 years, occasionally reaching temperatures higher than 400°C, and burning approximately six tons of ^{235}U in the process. So, what happened to the radioactive waste (and the associated decay products) that was produced? Most of it is still in Oklo, long after the reactor shut down. The large majority of the fission products stayed put. Plutonium, americium, neptunium, and curium didn't move. A few minor isotopes of molybdenum, silver, iodine, krypton, xenon, rubidium, cesium, strontium, cadmium and lead did move, sometimes locally, sometimes further (Brookins, 1978). Considering the fact that the Oklo ore deposit is riddled with fractures and joints (potential pathways for fluid and radionuclide release) the inertness of the waste is impressive (Faure, 1991). It is difficult to imagine more compelling testimony to the ability of rocks to attenuate high level radioactive waste.

DNAPLS

Roughly 60% of Superfund sites are estimated to have DNAPLs. Recall from above that DNAPLs are dense hydrocarbon pools that often move contrary to the direction of groundwater flow. Instead they tend to sink through the water table, often following fractures or other heterogeneities (see Figure 7.13). If they're out of reach of the backhoes then chances are they will never be removed, and in many cases the time elapsed between detection and remediation is sufficiently long to assure that they can't be dug out, much less removed by pumping and treating the water. Consequently, DNAPLs remain at depth with the potential to bleed their more soluble components into solution at the DNAPL-groundwater interface. The low surface-to-volume ratio of DNAPLs quite often means that it will take decades to centuries for DNAPLs to dissolve into groundwater. In this way DNAPLs constitute what appears on environmental time scales to be a near-perpetual source of contamination that can't be touched. The track-record for *in situ* removal of water-soluble contaminants is not good (see below). For DNAPLs it is awful. The quantity of NAPL recovered by commonly used recovery techniques is typically a trivial fraction of the total NAPL available to contaminate groundwater, and mobile NAPL recovery typically recovers less than 10% of the total NAPL mass in a spill (Wiedemeier et al., 1996). Note that missing even a small amount of DNAPL mass means that re-contamination of the groundwater can occur. Freeze and Cherry (1989) noted that "at sites where

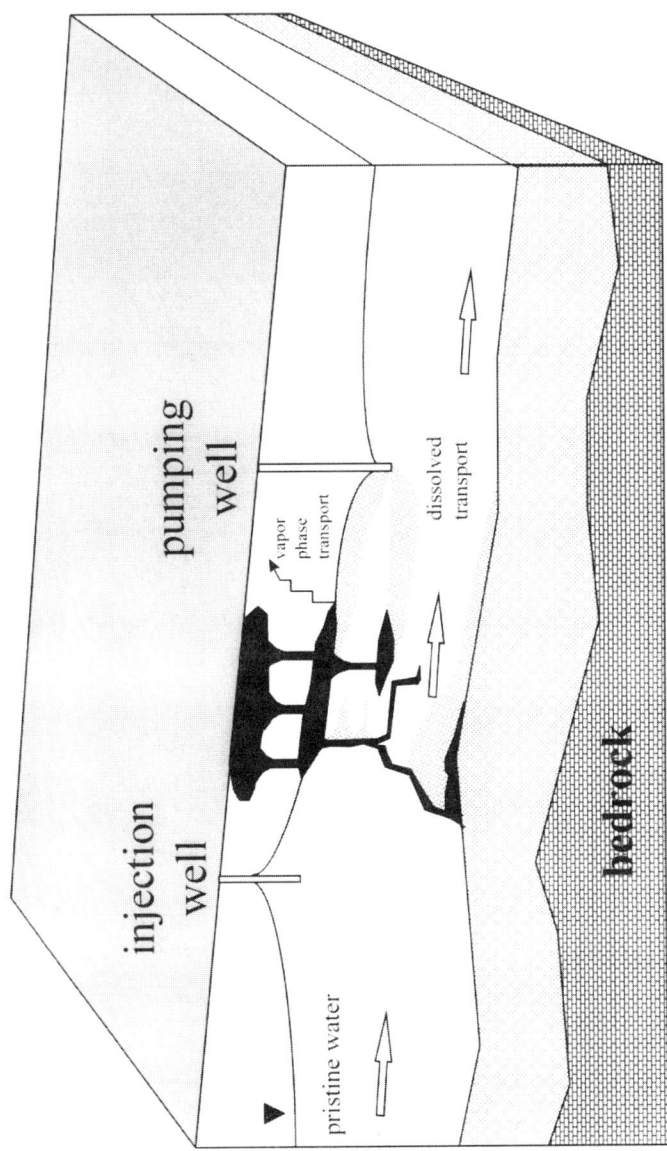

FIGURE 7.13 Schematic of pump-and-treat being used to treat DNAPL at depth.

DNAPLs are a problem, the local groundwater has terminal cancer. A cure in the form of returning the aquifer to drinking water standards is unachievable at any cost".

The primary threat comes about due to the constant bleeding of water-soluble DNAPL components into groundwater. Recall that the same components are often the molecules that are most rapidly biodegraded. Again, this depends on the availability of electron acceptors, nutrients, and/or cometabolites (see below). Other than this, there appears to be very little that can be done with DNAPLs. This has not prevented the expenditure of enormous amounts of money to treat and/or understand DNAPLs. The cost of DNAPL remediation attempts should always be compared against the cost of wellhead treatment and/or future use restrictions. It is exceedingly difficult to demonstrate that *in situ* treatment efforts have been successful on any scale that uses public health protection as a goal. At the same time a number of remediation approaches can make DNAPLs worse. Drilling, flushing/solubilizing, and dewatering coupled to vacuum extraction can move DNAPLs to places they would otherwise not go.

Two DNAPL components which are the target of the bulk of remedial efforts are the solvents TCE and PCE. Evidence is beginning to accumulate demonstrating natural attenuation of both (see Table 7.4).

A number of recent studies demonstrate natural attenuation of chlorinated organic contaminants in the field. Major et al. (1991) showed evidence for intrinsic bioremediation of up to 27 μM (1 μM = 10^{-6} mol) PCE to ethane and ethene at a chemical transfer facility in North Toronto. Methanol and acetate were found in the contaminated groundwater, and were thought to be the primary substrates for degradation. Fiorenza et al. (1994) demonstrated reductive dechlorination of 0.1 μM PCE, 11 μM TCE, 41 μM TCA, and 440 μM dichloromethane (DCM) contamination at a carpet-backing manufacturing plant in Hawkesbury, Ontario. High levels of fatty acids and methanol were thought to be the primary metabolic targets. Martin and Imbrigiotta (1994) estimated masses of TCE contamination at the Building 24 Site of the Picatinny Arsenal in Morris County, New Jersey. TCE biodegradation was calculated to be sizeable and was thought to occur through reductive dechlorination under reducing conditions. TCE degradation was modeled using as input microcosm studies, which may in this case have underestimated degradation (Martin and Imbrogiotta, 1994). Wilson et al. (1994b) documented anaerobic degradation of TCE at an NPL site in St. Joseph, Michigan. Calculated rates of TCE degradation were between 0.0076/week and 0.0024/week. McCarty and Reinhard (1993) presented evidence pointing to intrinsic bioremediation of chlorinated organics under conditions of methane fermentation in municipal landfillls. Lee et al. (1995) examined the DuPont Necco Park Landfill in Niagara Fall, New York, which had groundwaters contaminated with a variety of organic solvents (see Table 7.4) as well as metals.

Maximum levels were: 160 μM PCE; 1400 μM TCE, 140 μM cDCE, 48 μM VC, 77 μM TeCA, 110 μM 1,1,2-TCA, 7.0 μM 1,2-DCA, 59 μM CT, 670 μM CF, and 720 μM DCM (Lee et al., 1995). In one section (the F zone) Lee et al. (1995) calculated relatively rapid decreases of total volatile organic carbon (TVOC) corresponding to a biodegradation rate constant of 0.7/year, and a half-life of approximately one year. This is slightly lower than the half-live measured by Wilson et al. (1994b) at the St. Joseph, Michigan Site (Lee et al., 1995). The latter, for TCE, were between 1.7 and

TABLE 7.4
Organic Contaminants at Necco Park Site

Chloroethenes	Chloroethanes	Chloromethanes
Perchloroethene (PCE)	Hexachloroethane (HCA)	Carbon tetrachloride (CT)
Trichloroethene (TCE)	Pentachloroethane (PCA)	Chloroform (CF)
cis-1,2-Dichloroethene (CDCE)	Tetrachloroethane (TeCA)	Dichloromethane (DCM)
trans-1,2-Dichloroethene (TDCE)	1,1,1-Tetrachloroethane (1,1,1-TCA)	Chloromethane (CM)
1,1-Dichloroethene (1,1-DCE)	1,1,2-Trichloroethane (1,1,2-TCA)	
Vinyl chloride (VC)	1,1-Dichloroethane (1,1-DCA)	
	1,2-Dichloroethane (1,2-DCA)	
	Chloroethane (CA)	

From Lee, M. D. et al., in *Intrinsic Bioremediation,* R. E. Hinchee et al., Eds. pp. 205–222, Battelle Press, Columbus, OH, 1995. With permission.

TABLE 7.5
Field Degradation Rates of Chlorinated Organics

Reaction	$t_{1/2}$ (in years)	Number of sites
PCE $\Rightarrow$ TCE	1.2	7
TCE $\Rightarrow$ DCE	1.19	15
DCE $\Rightarrow$ VC	1.05	12
VC $\Rightarrow$ Eth	1.22	9

From Buchanan, 1996.

5.8 years. Weaver et al. (Weaver et al., 1996) calculated net apparent degradation rate constants for TCE (0.3 — 1.7 yr^{-1}), *cis*-DCE (0.54 — 4.0 yr^{-1}), and vinyl chloride (2.6 — 20 yr^{-1}) across a number of transects at the St. Joseph site. Buchanan (1996) presented a compilation of measured degradation half-lives for chlorinated solvents at 43 sites (see Table 7.5) which fall in the range measured at St. Joseph as well.

Cox et al. (1995) demonstrated natural biodegradation of TCE and 1,1,1-TCA in the groundwater near a once unlined septic-waste lagoon near Sacramento, California. TCE and 1,1,1-TCA were found to degrade by two paths. One resulted in complete anaerobic degradation of TCE and 1,1,1-TCA to non-chlorinated breakdown products. The second involved breakdown to vinyl chloride and chloroethane under anaerobic conditions, followed by the latter's subsequent aerobic degradation. Nevertheless, the respective breakdown rates were insufficient to control the migration of contaminants at the site. Guest et al. (1995) examined three different sites at an Air Force base in Colorado that were contaminated by chlorinated organics.

Degradation was occurring at two of the sites, but not at the third. The latter appeared to be the most oxidizing of the three sites, pointing up the importance of reducing conditions for reductive dechlorination. Bradley and Chapelle (1996) demonstrated that vinyl chloride can be oxidized in soils by electron acceptors other than free oxygen. Specifically, Fe(III) can be used as an electron acceptor by soil microorganisms to degrade vinyl chloride, particularly if the trivalent iron is in a bioavailable form (Bradley and Chapelle used EDTA-complexed Fe^{3+}). Hale et al. (1996) examined degradation of TCE at a site where there were high levels of BTEX compounds, ammonia, and dissolved oxygen. The presence of *cis*-1,2-DCE, elevated chloride concentrations, and aerobic TCE-degraders pointed toward aerobic degradation of TCE.

From the above it is clear that the redox sequence a plume of PCE or TCE encounters on its trip from source area to receptor largely determines the dose delivered at the latter. It is consequently critical for a natural attenuation site assessment that the redox zonation of a plume be understood in at least a qualitative sense (see e.g., Chapelle, 1996). Under anaerobic conditions this is probably best done by monitoring H_2 levels, and by tracking the changes in soil levels of nitrates, sulfates, iron, and methane (see e.g., Chapelle, 1996).

It is likely that evidence for natural attenuation of chlorinated organic components of DNAPLs will continue to accumulate. Natural attenuation has already been chosen as the final corrective action at two recent Superfund sites containing chlorinated aliphatics, including TCE: West Plume, St. Joseph Site, in St. Joseph, Michigan; and West Management Unit Plume, Dover Air Force Base, Dover, Delaware. The EPA's study of natural attenuation at the St. Joseph site can be downloaded from the world wide web address:

http://www.epa.gov/ada/stjoseph.html

TNT

As noted earlier, TNT sorbs to soil minerals and can be biodegraded. Generally both processes are more effective under reducing conditions. Figure 7.14 shows a proposed breakdown pathway for TNT. For the pathway in Figure 7.14, breakdown might be tracked by the appearance of products having progressively fewer nitro groups, and ultimately toluene, which is broken down and used in the tricarboxylic acid cycle. Although soil microorganisms are able to break down TNT, it is not their preferred substrate (Bradley and Chapelle, 1995). Bradley and Chapelle showed, using soil microcosms containing TNT-contaminated soils from a site in Missouri, that cyclic wetting and drying might concentrate TNT to the point where it would inhibit microbial breakdown. Cool temperatures and high clay contents are also thought to work against rapid degradation (Funk et al., 1995). Composting of TNT causes a conversion to the solvent extractable products 4-amino-2,6-dinitrotoluene and 2-amino-4,6-dinitrotoluene and/or attachment to soil organic matter (cellulosic material + humin + humic acid + fulvic acid) (Pennington et al., 1995). Uptake and transformation of TNT by aquatic plants has been demonstrated by Hughes et al. (1997).

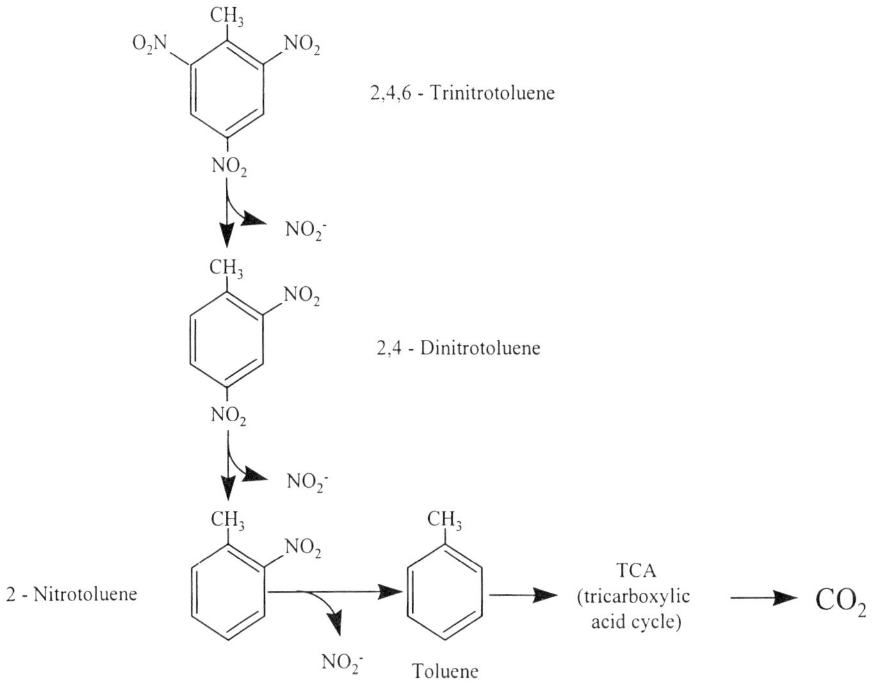

Case Studies

FIGURE 7.14 TNT breakdown. (After Duque et al., 1993).

The first regulator-sanctioned use of natural attenuation for closure of a site with explosives-contaminated groundwater was signed in early 1996. The decision covered the TNT leaching beds of the Sierra Army Depot near Herlong, California. The plume is 70 feet below ground, covers roughly 28 acres, and consists of explosives residues as well as TCE. A number of conventional cleanup procedures were considered, including pump-and-treat, ultraviolet oxidation, and treatment with granular-activated carbon filters. Natural attenuation was ultimately chosen, partially based on the fact that it cost between 1/6th and 1/10th of the standard approaches. Long-term monitoring of the plume and institutional control are critical components of the site closure plan. Drinking water supplies were 2 miles away, and the Army agreed to implement pump-and-treat, followed by filtering if the plume becomes a threat.

HEXAVALENT CHROMIUM

Chromium is one of the most common metal contaminants in soils, and, due to its anionic state (CrO_4^{2-}) can potentially be transported readily. Recall from Chapter 3 that when hexavalent chromium is reduced to its trivalent state it is both less toxic, and insoluble. In general, pump-and-treat methods can remove a substantial fraction of chromate from soils. Nevertheless, extraction efficiencies drop over time, and a significant fraction often remains at depth (Palmer and Puls, 1994; Puls, 1995). Removing this last fraction, consequently, becomes quite expensive. The extra expense may be unnecessary if it can be shown that the latter fraction is naturally attenuated through reduction to Cr(III). Demonstrating natural attenuation of hexavalent chromium in soils often is a matter of showing that there are sufficient electron donors at depth to cause the transformation, and that rates of chromate reduction are sufficiently rapid so as to minimize chromate levels at a point of complicance. Electron sources include (Palmer and Puls, 1994): aqueous Fe^{2+}, Fe^{2+} at mineral surfaces (e.g., pyrite, Fe(II)-silicates), and soil organic carbon. Because a contaminant plume is likely to displace the existing groundwater, and because diffusion is slow, the aqueous electron donors (Fe^{2+}) are probably less important. In fact, less than 1% of the reducing capacity of soil has been calculated to come from the aqueous phase (Palmer and Puls, 1994). The solid phase accounts for the remainder.

For natural attenuation of hexavalent chromium it must specifically be demonstrated that (Palmer and Puls, 1994):

1. There are natural reductants present within the aquifer;
2. The amount of Cr(VI) and other reactive constituents does not exceed the capacity of the aquifer to reduce them;
3. The time scale required to achieve the reduction of Cr(VI) to the target concentration is less than the time scale for the transport of the aqueous Cr(VI) from the source area to the point of compliance;
4. The Cr(III) will remain immobile; and
5. There is no net oxidation of Cr(III) to Cr(VI).

Apparently, the most difficult of the above to obtain are the rates of reduction and back-reaction (oxidation) (Palmer and Puls, 1994). Moreover, the occasional sorption and formation of minerals containing CrO_4^{2-} sometimes makes it difficult to demonstrate unambiguously that decreases in groundwater Cr(VI) can be completely be ascribed to reduction. Cr(III) can be oxidized back to hexavalent chromium by dissolved oxygen and manganese dioxide particles (Eary and Rai, 1987).

The mass of hexavalent chromium in the subsurface can be estimated by leaching "exchangeable" chromate with a potassium phosphate solution (James and Bartlett, 1983), and by subsequently estimating how much of the residual remains in the form of solid $BaCrO_4$. The latter is done by the methods cited in Palmer and Puls (1994). The mass of hexavalent chromium in soils, $Cr(VI)_{tot}$, is then (Palmer and Puls, 1994):

$$Cr(VI)_{tot} = [CrO_4^{2-}] + 1000\rho_b[Cr(VI)_{exchangeable} + Cr(VI)_{BaCrO4}]/\theta$$

where ρ_b is the dry bulk density of the soil (g/cm^3) and θ is porosity. Integration of this expression over the volume of contaminated soil is then the total amount of hexavalent chromium.

The presence of Cr(III) in soils indicates that reduction of hexavalent chromium is occurring. The amount of Cr(III) attached to the soil matrix can be ascertained by dissolving Cr, Fe, and Al hydroxides with a dithionate-citrate-bicarbonate (DCB) leach. Unfortunately, the latter also dissolves sparingly soluble Cr(VI) solids, which makes the mass balance more ambiguous. Bartlett and James (1988) have outlined a number of approaches to predicting the capacity of an aquifer to reduce hexavalent chromium, or, alternatively, oxidize trivalent chromium. One method involves measuring the total amount of soil organic carbon by reacting 2 to 3 grams of uncontaminated soil with a mixture of 1 normal $K_2Cr_2O_7$ and 1 normal H_2SO_4 for half an hour. After reaction, the total amount of Cr(VI) remaining in solution is measured, and the amount reduced by the soil organic carbon calculated by difference. This acid-rich digestion may overestimate the reducing capacity of the soil (Palmer and Puls, 1994), consequently, less corrosive tests have been used as well. $K_2Cr_2O_7$ added to a 0.01M H_3PO_4 pH-buffered solution and allowed to react for 18 hours is thought to give a good measure of the reductive capacity of soils (the phosphate buffers the pH to near neutral while competing with chromate for adsorption sites). Oxidation of trivalent chromium can be roughly estimated by adding 2.5 grams of moist soil to a 1mM $CrCl_3$ solution, shaking for 15 minutes, and then adding a $KH_2PO_4^-K_2HPO_4$ solution, from which the supernatant is centrifuged off, and filtered. The solution is then analyzed for Cr(VI), which corresponds to the amount of trivalent chromium oxidized by the soil.

Palmer and Puls (1994) note that it is very difficult to estimate hexavalent chromium reduction solely on the basis of monitoring well results. Because hexavalent chromium can form precipitates, sorb to the aquifer matrix, and be reduced, the actual contribution of the latter process to contaminant level decreases often cannot be unambiguously ascertained. One of the most successful attempts at quantifying natural attenuation of hexavalent chromium is that of Henderson (1994), who demonstrated reduction of hexavalent chromium in the Trinity Sand Aquifer of west Texas. Hexavalent chromium from a number of chrome plating operations seeped into the aquifer between 1969 and 1978. Maps of the resulting plume (1986 and 1991) were analyzed to show that nearly three quarters of the hexavalent chromium initially present had been removed. K_d's were measured on the aquifer material, callibrated against observed plume movement, and, subsequently used with a transport model and integrated contaminant masses to calculate a first order reduction rate for hexavalent chromium reduction of 3.2×10^{-5} per hour. The corresponding half-life for reduction of hexavalent chromium was determined to be 2.5 years, and the calculated rate constant is very close to what is predicted from laboratory measurements. Fe^{2+} and aquifer organic matter were thought to be the primary reducing agents. The primary sink for chromium was thought to be the formation of $Cr(OH)_3$. Using best estimates for the remaining hydrologic inputs, Henderson (1994) calculated that contaminant levels would decrease through natural attenuation to below MCLs within a decade.

PAHs

PAH-contaminated soils and groundwaters show up near manufactured-gas plants (MGPs), where coal tar was made, and distilled to form creosote, a wood preservative. Recall from Table 6.3 that PAHs are generally observed to degrade under aerobic conditions. In the field, degradation rates are often seen to vary widely from PAH to PAH. Artificial plumes of m-xylene, napthalene, and dibenzofuran emanating from a residual coal tar source were observed to behave quite differently over a 3-year time span (King et al., 1995). m-xylene reached a maximum plume distance and mass, and then began to recede and lose mass. Dibenzofuran reached a steady-state, where biodegradation matched input from the source. Napthalene was observed to continue to increase in mass and plume extent (King et al., 1995). Durant et al. (1995b) demonstrated that indigenous PAH-degrading microorganisms were probably attenuating a plume below a former MGP site. The primary evidence was degradation of ^{14}C-labeled napthalene which had been added to soils collected from sites which had low levels of contamination. Subsequent efforts in the same vein by Durant et al. (1995a) showed that aerobic degradation proceeded even under low temperatures (10°C) and low nutrient (phosphate, nitrate) availability. Jones (1996b) examined the long-term evolution of a plume downgradient from a DNAPL made up of coal tar from an MGP, which had been deposited in a trench beginning 30 years ago. The source material was removed in 1991. Of the constituents remaining in the plume those observed to have migrated farthest in the interim were the low molecular weight aromatics, which were subsequently observed to have been most effectively degraded by indigenous microorganisms.

BIOAVAILABILITY OF LEAD AND ARSENIC

Risk assessments for As-contaminated sites assume that soil As is completely bioavailable (Davis et al., 1996). Recall from Chapter 5 that, for a contaminant solid to be available to groundwater, it must be both soluble in that particular water and must dissolve rapidly as well. For a contaminant solid to be bioavailable the same requirements exist, though the fluid is no longer a groundwater. If solids are inhaled (e.g., asbestos), their residence time in the lung depends on their solubility in, and dissolution into, a fluid quite different in composition from normal groundwaters (see Hume and Rimstidt, 1992). If metal-containing solids are ingested, the attendant metal release depends on the solid solubility, dissolution rates, and residence time of the mineral in acid stomach fluids, between pH 1.3 (fasting) and 2.5 (average) and 4.0 (after eating). The small intestine has a pH of around 7 (Ruby et al., 1996). Ruby et al. (1996) showed that As-containing solids (metal-arsenic oxides and phosphates) likely to be ingested by residents exposed to Anaconda, Montana soils are sparingly soluble in the digestive tract. Moreover, their dissolution into the latter is hindered further by the formation of carbonate and silicate rinds which tend to coat the contaminant-bearing phases. Ruby et al. (1996) proposed this as an explanation for the apparently low bioavailability of arsenic in an otherwise As-rich environment. There is also considerable evidence that lead bioavailability depends on the identity of the solid it is contained in as well (Rieuwerts and Farago, 1995; Ruby et al., 1996)

ACID MINE DRAINAGE

Acid mine drainage (AMD) presents a special problem. Fluxes of acidity are large, and associated metal movement is non-trivial. Because the volume of tailings is enormous, only rarely can the source of both the acidity and metals be removed easily. When atmospheric oxygen comes into contact with ore bodies or mine tailings containing sulfide minerals such as pyrite (FeS_2, "fool's gold") the chemical reaction which ensues is:

$$FeS_2 + 7/2O_2 + H_2O \Leftrightarrow Fe^{2+} + 2SO_4^{2-} + 2H^+$$

The reaction above says that acidity (H^+) is produced when minerals containing sulfide (S^-) come into contact with water and oxygen at the Earth's surface. The elevated H^+ (i.e., low pH), and the other liberated heavy metals (e.g., zinc, cadmium, arsenic, and lead, which are present as impurities) are threats to aquatic organisms. At the same time their release is almost unavoidable when water and oxygen seep through mine tailings and abandoned mine shafts.

It should be pointed out that metal sulfides have been dissolving into natural waters for billions of years (There is even some evidence that Life began on metal sulfide surfaces (Wachtershauser, 1988). However, the excess acidity has been taken up by reactions with carbonate and silicate minerals for just as long. There are two ways to minimize the impacts of AMD: (1) Somehow stop metal sulfide dissolution by preventing the access of oxygen and/or water, or by adding a chemical agent to inhibit reaction at the mineral surface, and/or (2) accelerate the neutralization reaction by bringing minerals which buffer AMD into close proximity.

The slowing down of reactions by sorbed reactants is a common and natural process. The dissolution (and growth) of many soil and marine minerals depend on what molecules are sorbed to their surfaces (see e.g., Brady and House, 1996), and in theory, it is possible to "engineer" reaction rates to be slower (or faster) by tinkering with the chemistry of the reaction. For example, phosphate is often used to inhibit the appearance of $CaCO_3$ "scale" which can clog pipes. Phosphate has likewise been observed to adsorb to pyrite and "choke-off" dissolution (e.g., Huang and Evangelou, 1994). It has therefore been suggested that AMD in some cases might be alleviated by injecting phosphate as a leach solution.

One approach to remediating AMD once dissolution has occurred and acid released is to use anoxic limestone drains (ALD's) (Hedin et al., 1994). Basically, AMD is channeled into a ditch which has been filled with limestone cobbles and chips covered with a soil cap. The limestone dissolves slowly when wetted by the AMD and adds alkalinity (acid-buffering capacity) to the water. The soil cover maintains a semi-reducing environment to prevent the precipitation of trivalent iron (hydr)oxides, which might otherwise clog the system. The latter may precipitate out at the end of the drain and possibly scavenge some of the heavy metal contamination out of solution. This leads to a drop in pH due to the combining of hydroxyls from solution with the iron that is conteracted by conversion of bicarbonate to CO_2.

In effect, the engineered approach of ALDs optimizes the acid neutralization that happens naturally when AMD encounters base-containing carbonates and silicate minerals, just as phosphate inhibition of pyrite dissolution might occur in nature as well. These two examples are useful in that they suggest that an understanding of natural attenuation processes might give clues to effective engineered attenuation processes. Another example of this is the use of phosphate amendments to sequester, and render non-bioavailable, lead in contaminated soils (see e,g, Ruby et al., 1994). Lead phosphates are often the most stable (insoluble) form of lead in soils. Ruby et al. (1994) demonstrated that lead phosphate minerals were present at a port facility where the soils were contaminated by the transshipping of galena (PbS) ore, and smelting dross. Airborne phosphorus from an adjacent phosphoric acid plant combined with the lead to form a number of lead phosphate minerals. Thermodynamic calculations suggest that the latter are exceedingly insoluble, and able to limit lead release to groundwaters, as well as to acidic stomach fluids, if ingested. In effect, combination of lead with phosphate renders it non-bioavailable. Phosphate soil amendments might therefore provide long-term isolation of lead in contaminated soils. Note that phosphate sorbs or forms insoluble solids with a number of metals of environmental concern (Laperche et al., 1996). Valsami-Jones et al. (1995) observed exchange of UO_2^{+2}, $Th(OH)_4$, and Pb^{2+} for apatite Ca^{2+}, and the subsequent formation of insoluble lead phosphate and calcium uranyl-phosphate. Consequently, phosphate amendments, or phosphate mineral backfills have been considered for limiting the transport of a number of metals other than lead, as well as for the long-term isolation of radionuclides. Similarly, treatment of acid soils with carbonate has been shown to be an effective means for isolating radioactive strontium (Browman and Spalding, 1984), presumably as a trace component of newly formed carbonate minerals.

In the past the response to contaminated soil or groundwater has been to: (1) dig it up, or (2) pump it out (or at least try to). Both of these are expensive. The first one moves the problem somewhere else. The second one generally doesn't even get the bulk of the contaminant out of the ground. The AMD-ALD and phosphate amendment examples above suggest that natural processes often point to potential long-term remedies.

8 Demonstrating Natural Attenuation

There has been a historical prejudice against the reliance on natural attenuation for site remediation because the latter is perceived to be a "no-action" or "walk-away" solution to environmental contamination. In fact, implementation of natural attenuation for most sites is proactive, requiring numerous lines of biogeochemical and hydrologic evidence. Moreover, long-term monitoring is often a critical feature of site closure plan. The various approaches which have been proposed and/or followed to gain regulatory approval are outlined below.

FUEL HYDROCARBONS

The most comprehensive cleanup protocol that includes consideration of natural attenuation is that of Wiedemeier et al. (1995b) for the cleanup of fuel hydrocarbons at DOD facilities. The protocol of Wiedemeier et al. (1995b) is also a reasonable starting point for developing a realistic approach to site closure for other contaminants (e.g., chlorinated hydrocarbons, metals, radionuclides) as well. The technical protocol of Wiedemeier et al. (1995b) is outlined schematically in Figure 8.1. Natural attenuation is considered as a potential remedy at the very first step of the process, and many of the most critical decision points rely on a conceptual understanding of contaminant degradation in the subsurface. What is new and different is that the dynamics of contaminant attenuation take on a much greater role in site characterization and choosing of remedial alternatives.

Wiedemeier et al. (1995b) condense the protocol into an eight-step approach:

1. Review available site data,.
2. Develop a preliminary conceptual model and assess the potential for intrinsic remediation,.
3. If intrinsic remediation is selected as potentially appropriate, perform site characterization in support of intrinsic remediation.
4. Refine the conceptual model based on site characterization data, complete pre-modeling calculations, and document indicators of intrinsic remediation.
5. Simulate intrinsic remediation using analytical or numerical solute fate and transport codes that allow incorporation of a biodegradation term, as necessary.
6. Conduct an exposure pathways analysis.
7. If intrinsic remediation alone is acceptable, prepare long-term monitoring (LTM) plan.
8. Present findings to regulatory agencies and obtain approval for the intrinsic remediation with the LTM option.

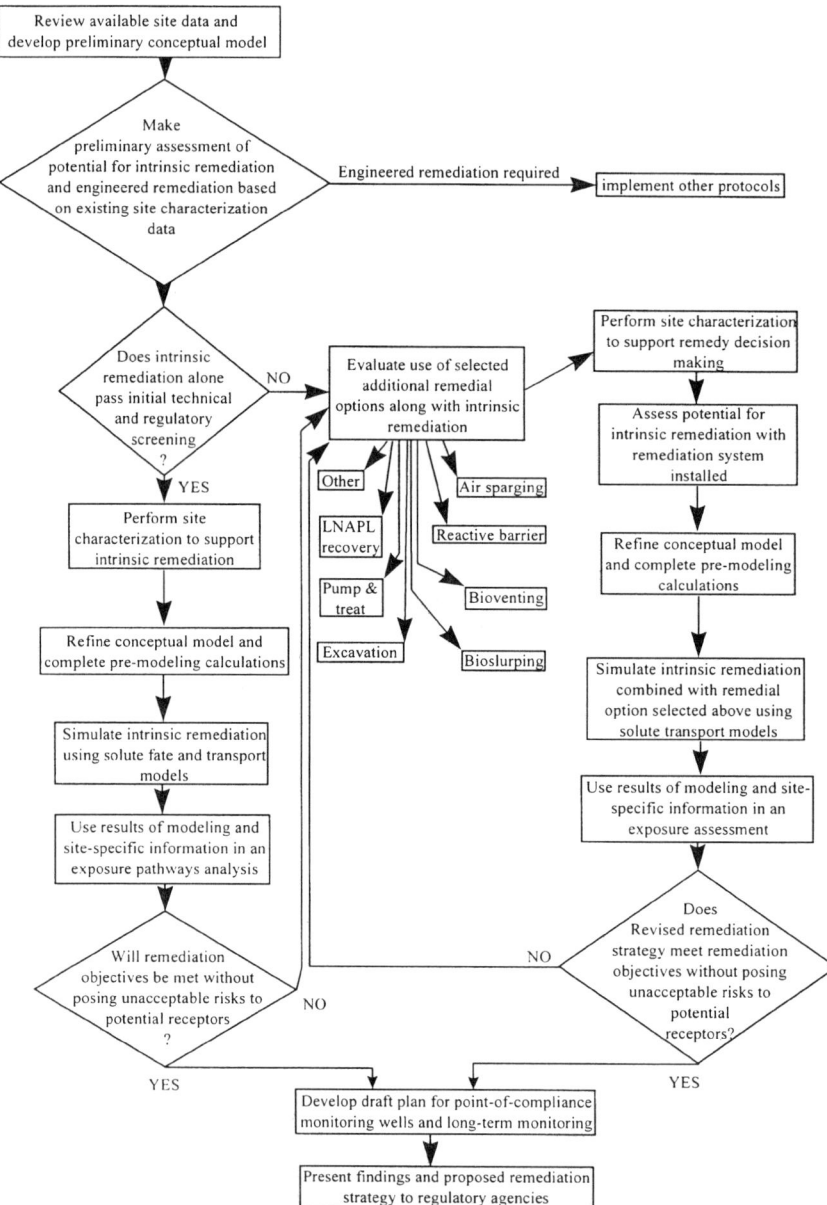

FIGURE 8.1 Schematic of fuel hydrocarbon natural attenuation protocol of Wiedemeier, T. H. et al., in *Intrinsic Bioremediation,* R. E. Hinchee et al., Eds., pp. 31–52, Battelle Press, Columbus, OH, 1995. With permission.)

Natural attenuation and intrinsic remediation are synonymous. Step 1, the review of all available site data, should be done with a mind to supporting natural attenuation as the remedial option. This will lead to a preliminary conceptual model (Step 2) which might, in turn, guide subsequent site characterization to fill in data gaps. The preliminary conceptual model is not a computer model, but a "ballpark" understanding of the site. Obvious data which should be considered in Step 1 include (Wiedemeier et al., 1995b): the nature and extent of the contamination; the timing of the release; and the nature of the contaminant. The nature of the contaminant is particularly important and a preliminary assessment of natural attenuation pathways based on the history of similar sites should be made early on. Standard hydrologic data (potentiometric surface, flow direction, hydraulic conductivity, flow rates, and lithology) should also be assessed in Step 1. An initial determination of pathway and distance to the most likely receptor should also be made. The development of the preliminary conceptual model requires the integration of the contaminant history, the assessment of biogeochemical attenuation, the site hydrology data, and the exposure determination.

Step 3 involves filling in the data gaps identified in Steps 1 and 2. In other words, construction of more than a preliminary conceptual model will more than likely be undoable because critical data (e.g., hydraulic conductivities, plume location and geometry) do not exist. Step 3 entails the collection of these data, particularly those which track natural attenuation in the subsurface. Obvious examples of the latter include electron acceptor (and/or methane) levels (e.g., O_2, NO_3^-, Fe^{III}) pointing to the oxidation of fuel hydrocarbons. Step 4 is the refining of the conceptual model to reflect the filling of the data gaps.

To document natural attenuation, Wiedemeier et al. (1995b) highlight the usefulness of contaminant contour maps, electron acceptor, metabolic by-product, and alkalinity maps. For subsequent refinement of the conceptual model, sorption must be estimated. For the non-polar organic contaminants this can be done using the methods outlined in Chapter 6.

To make predictions using biogeochemical models requires rate constants for the relevant breakdown reactions. In other words, it is not sufficient to make the case that biodegradation is occurring at depth. Actual rates are needed to estimate the ultimate lifetime of a plume. Table 8.1 compiles a large number of fuel hydrocarbon biodegradation rate constants measured in the lab and the field.

The measurement of degradation rates in the laboratory involves monitoring the breakdown of the contaminant of interest over time in a microcosm, and is fairly straightforward. Estimating a degradation rate in the field can be done by a number of different techniques. One can assume that a plume source (NAPL) bleeds off soluble components to remain at equilibrium with groundwater flowing past. In this case a steady-state flux of contaminant can be calculated as the product of the flow rate (liters/m^2/day), the cross-sectional area of the zone of contamination (m^2), and the solubility of the contaminant (mol/l), to give a degradation rate in mol/day. If the plume is at steady-state (neither advancing or retreating), the calculated degradation rate is equivalent to the actual degradation rate. If the plume is receding, the

TABLE 8.1
Organic Degradation Rates

Site	Contaminant	V (m/d)	Field results (mg/day)	Lab results (mg/day)
Borden, Ontario (Barker et al., 1987)	BTX stock solution injected into uncontaminated aquifer	0.09	benz. = 30 tol. = 37 m-xyl = 47 p-xyl = 55 o-xyl = 33	benz. = 58 tol. = 61 m-xyl = 50 p-xyl = 65 o-xyl = 54
Rocky Point, North Carolina (Borden et al., 1994)	Residual gasoline from UST	0.08	benz. = 0.0002 tol. = 0.002 1e-benz. = 0.0015 m,p-xyl. = 0.0013 o-xyl. = 0.0021	benz. = 0.0002 tol. = 0.002 1e-benz. = 0.0015 m,p-xyl. = 0.0013 o-xyl. = 0.0021
Kalkaska, Michigan (Chiang et al., 1989)	Natural gas condensate- BTEX	0.2	benz. = 0.007 p-xyl = 0.0107 napthalene = 0.0064	BTX = 0.01–0.1
Columbus, MS (MacIntyre et al., 1993)	Stock solution of benzene, p-xylene, napthalene, o-dichlorobenzene		o-DCB = 0.0046	
Sleeping Bear, Michigan (Wilson et al., 1994a)	Residual gasoline from UST release BTEX	0.2	benz. = N.S. tol. = 0.02–0.07 e-benz. = 0.03–0.011 m-xyl = 0.004–0.014 p-xyl = 0.002–0.010 o-xyl = 0.004–0.011	benz. = N.S. tol. = 0.007–0.04 e-benz. = N.S m,p-xyl = N.S. o-xyl = N.S.

Demonstrating Natural Attenuation

Site	Contaminant	Value	Concentrations (mg/L)
Indian River, Florida (Kemblowski et al., 1987)	Gasoline from UST-BTEX	0–0.4	benz. = 0.02–0.2
Morgan Hill, California (Kemblowski et al., 1987)	Gasoline-BTEX	0.05	benz. = 0.0085
(Wilson et al., 1994a)	JP-4 jet fuel	1.3	benz. = 0.0035
			benz. = below detection
			tol. = 0.05–0.013
			e-benz. = 0.03–0.05
			m-xyl = 0.02–0.1
			p-xyl = 0.02–0.08
			o-xyl = 0.21
Hill AFB, Utah (Wiedemeier et al., 1995a)	JP-4 jet fuel	0.5	benz. = 0.03–0.09
			e-benz. = 0.01–0.08
			m-xyl = 0–0.03
			p-xyl = 0.01–0.03
			o-xyl = 0–0.02
Patrick AFB, Florida (Wiedemeier et al., 1995a)	Unleaded gasoline from UST	0.13	benz. = 0–0.004
			tol. = 0.0006–0.004
			e-benz. = 0.0001–0.004
			m-xyl = 0.0001–0.004
			p-xyl = 0.001–0.003
			o-xyl = 0.004–0.02
Fairfax, Virginia (Bushcheck et al., 1993)		0.015	benzene = 0.00055
			tol. = 0.00045
			e-benz. = 0.00045
			m,p,o-xyl = 0.004
San Francisco, California (Bushcheck et al., 1993)		0.03	benzene = 0.0028
			tol. = 0.0022
			e-benz. = 0.0033
			m,p,o-xyl = 0.0023

**TABLE 8.1 (continued)
Organic Degradation Rates**

Site	Contaminant	v (m/d)	Field results (mg/day)	Lab results (mg/day)
Alameda County, California (Bushcheck et al., 1993)	Gasoline-BTEX	0.01	benz. = 0.002 tol. = 0.0017 e-benz. = 0.002 m,p,o-xyl = 0.0023	
Elko County, Nevada (Bushcheck et al., 1993)	Gasoline-BTEX	0.04	benzene = 0.001	
Traverse City (Wilson et al., 1990)	Aviation gasoline from UST-BTEX	1.5	benzene = 0.001 tol. = 0.2 m,p,o-xyl = 0.004	Toluene and ethylbenzene rapidly degraded in denitrifying microcosms after 56 day lag period
Broward Co., Florida (Caldwell et al., 1992)	Gasoline from UST-BTEX and MTBE	0.1	BTEX = 0.00012	
Pensacola, Florida (Bekins et al., 1995)	Creosote-phenols	0.3–0.12	Selected phenols completely degraded over a 100 day travel time through methanogenic aquifer.	Selected phenols were completely degraded over 100–200 days in methanogenic microcosms
Bemidji, Minnesota (Baedecker et al., 1993)	Crude oil-BTEX	0.25	Toluene and o-xylene depleted over 20 meters (200 day travel time); benzene and e-benzene depleted over 100 meters. Downgradient migration was limited by mixing with uncontaminated water.	98% benzene loss in 125 days and 99% toluene loss in 45 days in anaerobic microcosms
Perth, Australia (Thierrin et al., 1993)	UST-BTEX	0.4	benzene = N.S. tol. = 0.006 e-benz. = 0.003 m,p-xyl = 0.004 o-xyl — 0.006 napthalene — 0.004	Anaerobic columns with 14 ppm sulfate benz. = N.S. tol. = 2.3 e-benz. = N.S. o-xyl = N.S.

Demonstrating Natural Attenuation

Site	Contaminants	Rate (kg/day or other)	Comments	Notes
Manufacturing Plant (Davis et al., 1994)	Benzene	0.16	benzene >0.01	Over 90% benzene loss after 77 days under methanogenic and sulfate-reducing conditions
Cliffs-Dow (Klecka et al., 1990)	Charcoal wastes, phenols, napthalene	0.2–0.46	All organics degraded within 100 m of the source	
Hill AFB-2 Utah (Dupont et al., 1994)	UST	0.14	benz. = 0.02 kg/day e-benz. = 0.06 kg/day p-xyl = 0.06 kg/day	All organics degraded in aerobic microcosms within 30–60 days.
Picatinny Arsenal, New Jersey (Martin and Imbrogiotta, 1994)	TCE, 1,1,1-trichloroethane and metals	0.3–1.0	Reductive dechlorination determined from plume locations	In anaerobic microcosms TCE = 0.0001–0.003
St Joseph, Michigan (Wilson et al., 1994b)	TCE from lagoon/dry wells	0.1	TCE = 0.001–0.003	
Finger Lakes, New York (Major et al., 1991)	TCE, acetone, methanol		Reductive dechlorination determined from plume locations	

Note: N.S., not significant.

From Rifai, H. S. et al., in *Intrinsic Bioremediation*, R. E. Hinchee et al., Eds., pp. 53–56. Battelle Press, Columbus, OH, 1995. With permission.

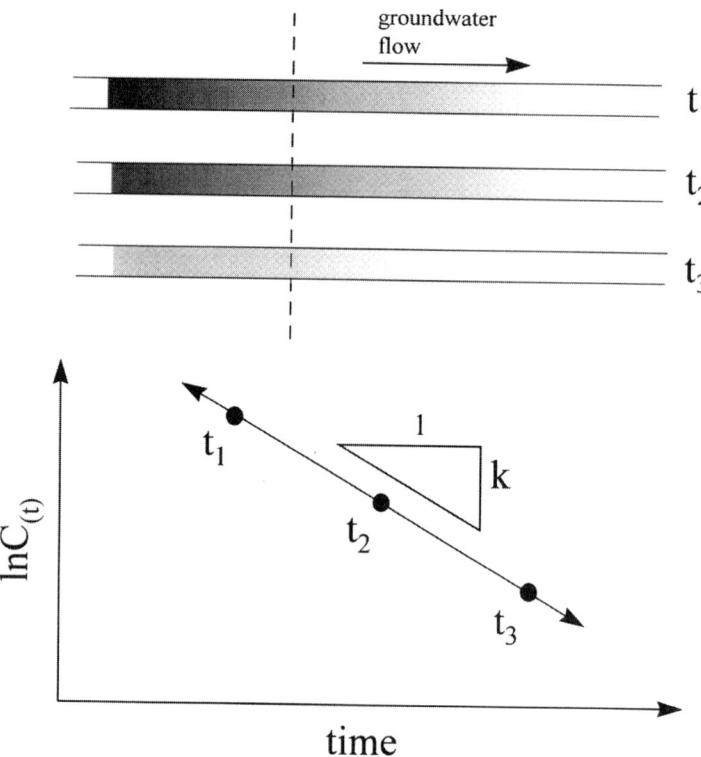

FIGURE 8.2 Calculation of biodegradation rate constant from changes in contaminant level.

calculated degradation rate is less than the actual degradation rate. If the plume is still expanding, the calculated degradation rate is greater than the real degradation rate.

A more precise approach, applicable to plumes of dissolved contaminants which shrink over time, assumes that all contaminant mass loss is due to biodegradation, and that contaminant levels do not vary vertically. In this model (Bushcheck and Alcantar, 1995):

$$C_{(t)} = C_{(t=0)} \exp[-kt]$$

where C is contaminant concentration (mol/l) at a given time, t (days), and k is the first order rate constant for degradation. Plotting the natural logarithm of $C_{(t)}$, measured at a given location over time, against time, should trace a straight line (if biodegradation is first order, and if the other assumptions are appropriate as well) having a slope of -k, the degradation rate constant (see Figure 8.2).

If the plume is stable, the approach of Kemblowski et al. (1987) may be followed, where x/v (= distance/velocity) is substituted in the equation above for time to give:

$$C_{(x)} = C_{(x=0)} \exp[-kx/v]$$

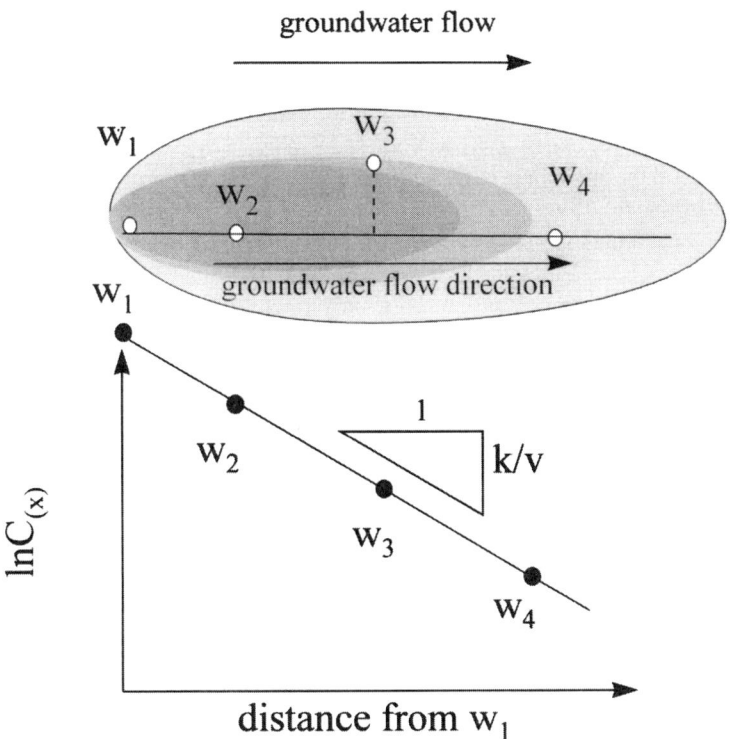

FIGURE 8.3 Calculation of biodegradation rate constant from change in plume position.

$C_{(x)}$ is the contaminant level at a given distance along the plume axis relative measured from the source, which is at x = 0. $C_{(x=0)}$ is the contaminant level at the source. A series of analyses done along the plume axis at a given time, when plotted as $\ln C_{(x)}$ against x, should define a line having a slope of $-k/v$ (see Figure 8.3). Multiplication of the latter term by the groundwater velocity, v, gives the rate constant. Note that both of the approaches above lump all contaminant loss pathways (sorption, dispersion, and biodegradation) into a degradation rate constant. If biodegradation is the dominant contaminant sink, the k's calculated above correspond to true, biodegradation rate constants. On the other hand, if sorption and/or dispersion are significant, the calculated k's will overestimate biodegradation. An alternative approach is to use an exact analytical solution to specifically determine a biodegradation rate from observations of plume morphology. Analytical solutions for one-dimensional transport which permit the estimation of advection, dispersion, sorption, and biodegradation are presented by Bushcheck and Alcantar (1995) and Domenico (1987). Perhaps the simplest method for extracting rate constants from field measurements may be to use a computer program, such as BIOSCREEN, to iteratively fit contaminant levels measured along the plume axis (for stable or receding plumes). BIOSCREEN will be discussed in greater detail below.

Step 5 of the AFCEE (Wiedemeier et al., 1996) protocol requires that plume behavior be predicted in space and time. Computationally this involves the solution of a number of differential equations describing fluid transport and chemical reaction. Numerous sources outline the mathematical underpinnings of reaction transport modeling — see for example; Huyakorn and Pinder, 1983, or Oelkers et al., 1996. The object of the modeling is to reproduce effectively the chemical and physical behavior of the site from a sufficiently fundamental understanding of the particular processes so that the long-term behavior of the plume might be confidently predicted. A host of site-specific factors, such as hydrologic and biogeochemical heterogeneities, guarantee that Step 5 will, at best, only roughly approximate reality. If field data have been collected over sufficiently long periods of time a model isn't needed to predict the future. Rarely is this the case though, and in the absence of data, a well-calibrated model is the next best substitute.

A model is only as good as its input. Required input includes (Wiedemeier et al., 1995b):

1. Hydraulic conductivity
2. Initial hydraulic head distribution
3. Flow direction and gradient
4. Effective porosity
5. Coefficient of hydrodynamic dispersion
6. Coefficient of retardation
7. Initial solute concentrations
8. Contaminant source concentration configuration, and rate of source decay/removal
9. Distribution and continuity of aquifer and aquitards
10. Groundwater recharge and discharge
11. Definition of physical and chemical boundary conditions
12. Rates of chemical reactions

The more site-specific are the input values, the more realistic will be the model output. Some of those above will be very difficult to approximate given the general quantity of knowledge at most sites (see e.g., Number 9). A variety of codes have been advanced to perform these calculations. Bioplume II (Rifai et al., 1988) has been used to argue successfully the natural attenuation option for fuel hydrocarbons at a number of sites. One of the more recent computer programs for this purpose is BIOSCREEN (Newell et al., 1996), which, in all likelihood, will be used with increasing frequency in the future to support remediation of organic contaminants by natural attenuation. It therefore deserves some further explanation. BIOSCREEN was generated through collaboration between AFCEE and the R. S. Kerr Environmental Research Center, and was intended to be used as a screening tool to specifically determine whether or not a full-scale evaluation of natural attenuation at a large site was warranted, or, in the case of smaller sites, as the primary evidence for natural attenuation (Newell et al., 1996). BIOSCREEN can be downloaded from the EPA web site at Ada, Oklahoma

Demonstrating Natural Attenuation

http://www.epa.gov/ada/kerrlab.html

The following summary largely follows the downloadable documentation. BIOSCREEN, given inputs of a number of hydrological and geochemical factors (see below), performs a series of calculations to specifically answer: (1) How far will a dissolved contaminant plume extend if no engineered remediation or source reduction is carried out? and (2) How long will the plume persist before natural attenuation processes cause it to dissipate?

The equation BIOSCREEN solves is Domenico's (1987) analytical expression for multi-dimensional transport of a decaying contaminant species. The primary assumptions of the latter are that: the aquifer and flow field are homogeneous and isotropic; molecular diffusion is minor and can be neglected; and adsorption can be treated with a linear isotherm (in essence, with a K_d). The spread sheet calculation requires as input hydrologic and geochemical data:

1. Seepage velocity — interstitial groundwater velocity
2. Hydraulic conductivity
3. Hydraulic gradient
4. Effective porosity
5. Longitudinal, transverse, and vertical dispersivity (BIOSCREEN uses published relationships to calculate these values, given a user estimate of plume length)
6. Retardation factor (If this is unknown, BIOSCREEN calculates it given input values of bulk density, f_{OC}, and, K_{OC})
7. First order decay coefficient (Alternatively, given a half-life $t_{1/2}$, the decay coefficient = $0.693/t_{1/2}$)
8. The respective differences in O_2, NO_3^-, Fe^{2+}, SO_4^{2-}, and CH_4 between background and the source area
9. The model area length and width
10. Simulation time
11. Source zone width, thickness, concentration, and soluble mass
12. Contaminant measurements from the field, to calibrate the model (if available)

Figure 8.4 shows a sample input that is fairly self-explanatory. BIOSCREEN calculates biodegradation in one of three ways; using a first-order decay constant (or, alternatively, a half-life); assuming instantaneous reaction with available electron donors; and not at all. In Figure 8.5 is shown the output calculated benzene plume size for the sample input in Figure 8.4, using half-lives of 0.01, 0.1, and 0.3 years after 5 years of reaction and transport of 100 kilograms of contamination. The calculation shown used as input the average electron donor levels from a large number of AFCEE sites. Once plume behavior is forward-modeled the output can be used to make remediation decisions.

Step 6 in the protocol of Wiedemeier et al. (1995b) simply requires that the most likely pathways for exposure be determined under current and reasonable future use

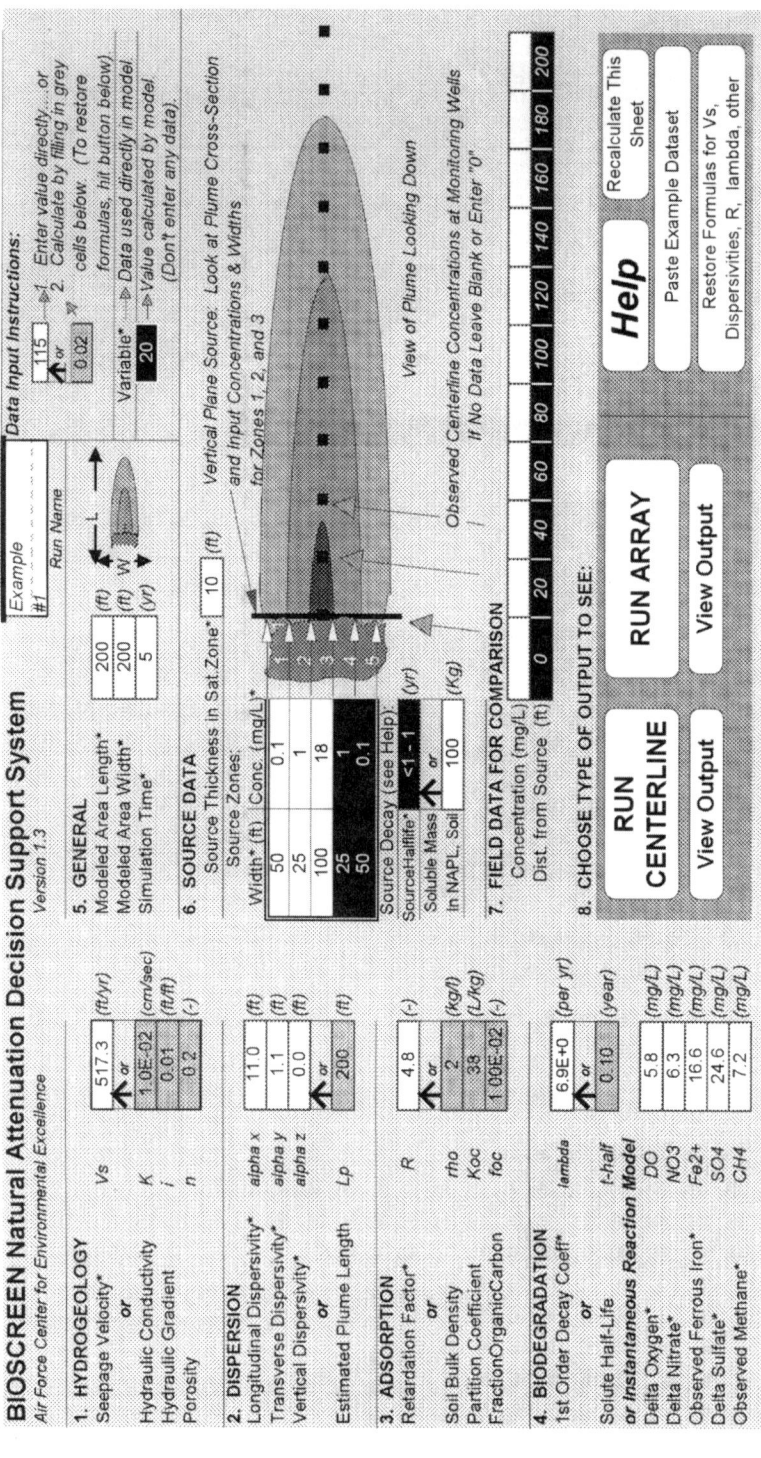

FIGURE 8.4 Example input for BIOSCREEN.

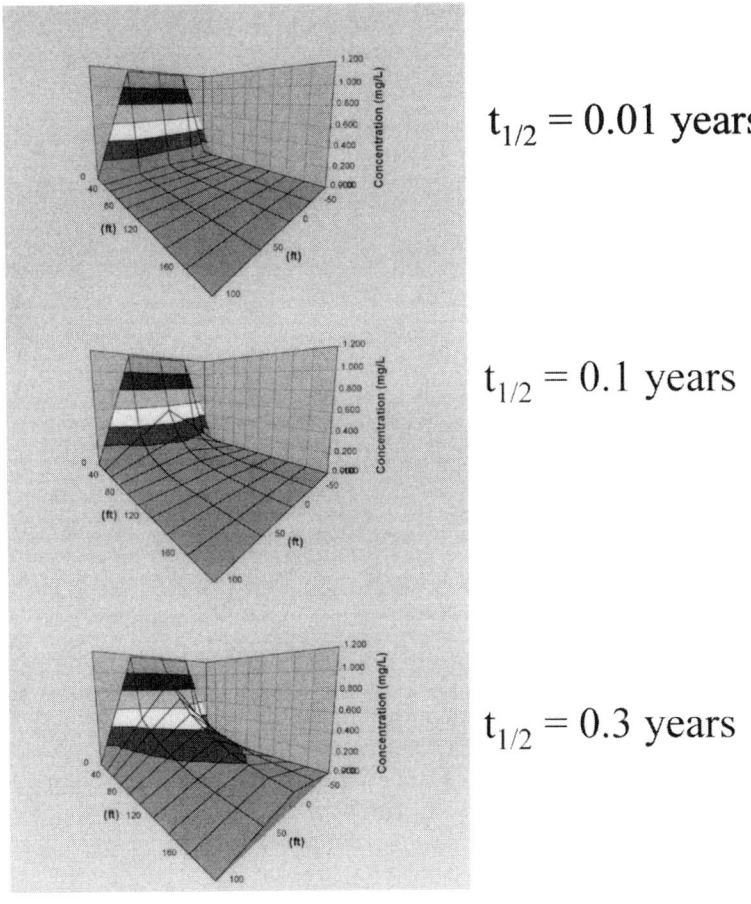

FIGURE 8.5 BIOSCREEN output for three biodegradation rates.

scenarios. If natural attenuation is demonstrably effective at minimizing exposure, a long-term monitoring plan should be formulated. The object of the latter is partly to provide a check on the possibility that uncertainties in the model and characterization have led to an overestimation of the efficacy of natural attenuation. Wiedemeier et al. (1995b) suggest two types of monitoring wells; long-term monitoring wells to determine if the behavior of the plume is changing; and point of compliance (POC) wells to detect contaminant movement outside the negotiated perimeter of containment. The latter might be used to trigger more active remediation, such as pump-and-treat, to contain the plume. Figure 8.6 illustrates the placement of the two types schematically. The LTM wells, by their function, should include analyses for contaminant levels, and if the latter are organic, likely electron acceptors (and/or methane) as well. POC wells, in the same situation, should sample for contaminant levels as well as dissolved oxygen (Wiedemeier et al., 1995b). Sampling frequency will depend on the trend seen in prior sampling. If predicted trends hold, there is less impetus for high-frequency sampling.

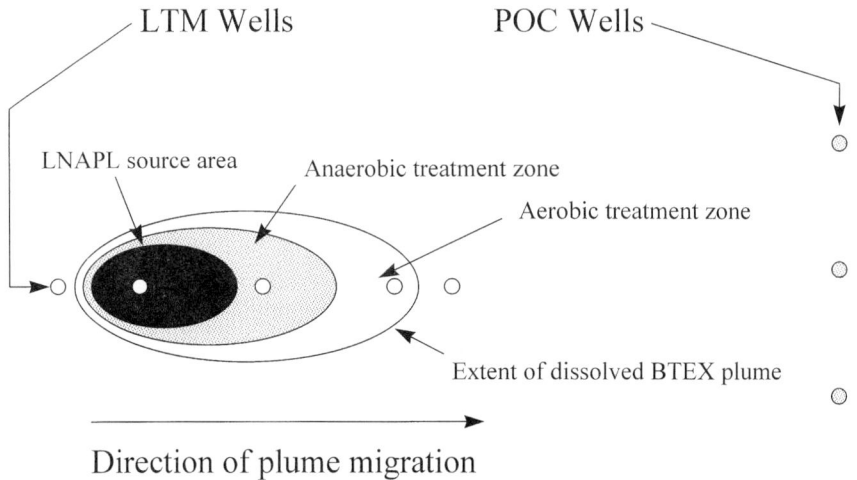

FIGURE 8.6 Long-term monitoring for natural attenuation. (From Wiedemeier, T. H. et al., 1995b.)

Step 8 in the Wiedemeier et al. (1995b) protocol, the negotiation with regulators, is critically important. As outlined elsewhere in this book, there appears to be an accelerating shift towards regulatory acceptance of natural attenuation. Nevertheless, in the absence of more fundamental programmatic change, regulators do not *have* to even consider natural attenuation. Quite simply, there is rarely enough field data to unequivocally demonstrate that natural attenuation alone will bring a site into compliance and models are, at best, less persuasive substitutes. For fuel hydrocarbons, the argument for natural attenuation relies primarily on three observations (Rifai et al., 1995a):

1. Compound disappearance. Obviously, the best evidence of all is if the contaminant of concern can be shown to be rapidly decreasing in concentration. Generally this requires the comparison of contaminant levels to the same degree for some non-reactive component (e.g., MTBE in a LUFT). Additionally, the differences in the rates tracer and contaminant plumes move can provide evidence of compound disappearance.
2. Loss of electron donors. Depressed electron donor (e.g., O_2, NO_3^-, SO_4^{2-}, etc.) levels, relative to a measured background, is good evidence that organic matter is being degraded.
3. Degradation products. Fe^{2+}, H_2S, CO_2, and methane are produced during the degradation of fuel hydrocarbons. Elevated levels relative to the background also point to degradation.

We believe that as field evidence for natural attenuation accumulates, the weight of evidence required for regulatory approval will shrink. If the natural attenuation alternative is examined at each site the field evidence should accumulate that much faster. For many sites (e.g., BTEX in non-conductive soils a long way from receptors) the effort and expense involved in computer modeling and long-term monitoring are probably not justified.

THE AFCEE PROTOCOL FOR CHLORINATED SOLVENTS

The overall approach of the AFCEE protocol emphasizes the iterative development of a conceptual model of natural attenuation, and stress is placed on assembling multiple lines of evidence for natural attenuation. The three lines of evidence used in the AFCEE protocol for natural attenuation of chlorinated aliphatic hydrocarbons are as follows:

1. Observed reduction in contaminant concentrations along the flow-path downgradient from the contaminant source
2. Documented loss of contaminant mass using geochemical data (e.g., loss of parent compound, appearance of daughter compounds, depletion of electron donors and acceptors, accumulation of metabolic by-products), or measured against the transport of a non-reactive tracer
3. Microbiological laboratory data attesting to biodegradation

According to Wiedemeier et al. (1996), at a minimum, the first and second, or first and third lines of evidence must be obtained. The second and third are used primarily for calculating biodegradation constants. Note that a reduction in contaminant levels downgradient might reflect any of a number of processes (adsorption, volatilization, dispersion, biodegradation); hence the appearance of daughter products, such as vinyl chloride, is a more definite evidence of biodegradation. By the same token, decreasing contaminant levels might indicate biodegradation if confirmatory evidence of biodegradation from microcosm studies are observed as well.

Wiedemeier et al. (1996), on the basis of the primary substrate (electron donor), identify four types of chlorinated solvent plumes — Types 1, 2, 3, and mixed. For Type 1 behavior, the primary substrate is anthropogenic carbon, such as BTEX compounds or landfill leachate. Under Type 2 conditions, conditions are generally reducing and there is rapid breakdown of highly chlorinated solvents. For Type 2, the primary substrate is native organic carbon, and breakdown of the more chlorine-rich organics is somewhat slower. In Type 3 plumes, there is little native or anthropogenic carbon, and the plume is relatively aerobic. In Type 3 plumes, breakdown of the highly chlorinated organics is minimal. Degradation of less chlorinated organics, which are more reduced, is more rapid though, assuming they are present in the first place. Single plumes can have areas which fall under each type. This is referred to as mixed behavior. Mixed behavior is important because often the juxtaposition of one type behavior against another hastens the sequential breakdown of oxidized, highly chlorinated compounds (PCE and TCE) to chlorine-poor, relatively reduced daughters (DCE and VC), to CO_2. Recall from Chapter 7 that the transition from chlorinated solvent to CO_2 often proceeds by way of reductions, followed by oxidations, hence the importance of sharp redox fronts. Type 1 or Type 2 behavior favors the first step. Type 3 behavior favors the last.

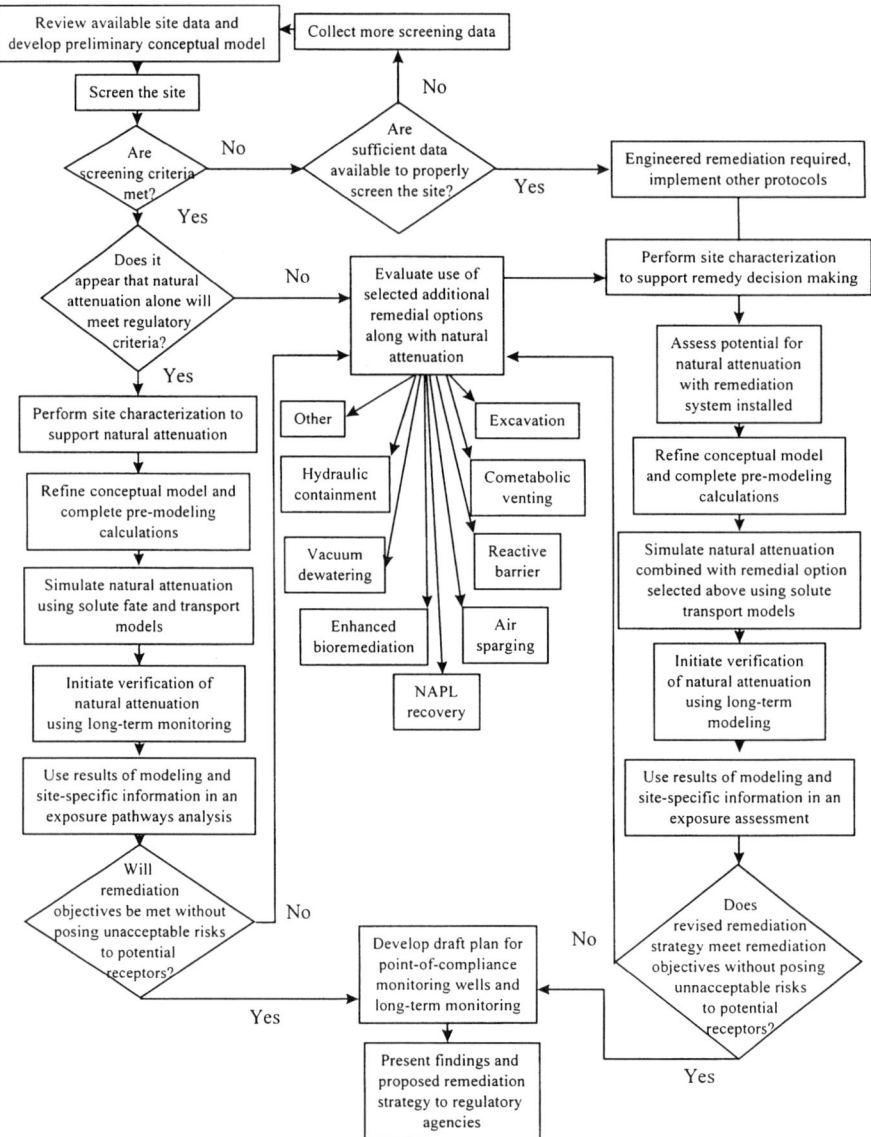

FIGURE 8.7 Schematic of chlorinated aliphatic natural attenuation protocol of Wiedemeier et al. (Wiedemeier et al., 1996).

The AFCEE protocol is shown in Figure 8.7. It is a 9-step approach which is similar in construct to the AFCEE fuel hydrocarbon protocol:

1. Review available site data, and develop a preliminary conceptual model
2. Screen the site and assess the potential for intrinsic remediation

3. Collect additional site characterization data to support natural attenuation, as required
4. Refine the conceptual model, complete pre-modeling calculations, and document indicators of natural attenuation
5. Simulate intrinsic remediation using analytical or numerical solute fate and transport codes that allow incorporation of a biodegradation term, as necessary
6. Identify potential receptors, and conduct an exposure pathways analysis
7. Evaluate the practicability and potential efficiency of supplemental source removal options
8. If natural attenuation alone is acceptable, prepare long-term monitoring (LTM) plan
9. Present findings to regulatory agencies and obtain approval for remediation by natural attenuation

Step 7, the evaluation of the practicability and potential efficiency of supplemental source removal, is the only step which does not have a corresponding number in the AFCEE fuel hydrocarbon protocol. Because of the correspondence between the two, there will be less emphasis on what each step entails, and more on the areas which are specifically different.

How to develop a conceptual model (Step 1) has been covered repeatedly above. Site screening (Step 2) is schematically detailed in Figure 8.8. Note that the collection of additional site characterization data (Step 3) during site screening is focused on determining if biodegradation is occurring. If it is, the remaining 5 steps revolve around figuring out: (1) how great a distance separates the contaminant from the nearest receptor; and (2) how rapidly the contaminant is being degraded along the way. The six steps of the screening procedure will be dealt with in turn.

Step 1. Determine whether biodegradation is occurring — this involves sampling at least six wells that are representative of the plume (The analytical protocol of Wiedemeier et al. (1996) is listed in Table 8.2).

Samples should be taken from the source; downgradient from the source, but still in the plume; downgradient from the plume; upgradient and laterally removed from the plume. Figure 8.9 shows the locations for sampling recommended by Wiedemeier et al. (1996). Sampling at the source defines the primary electron donor (e.g., BTEX or indigenous soil organics). Sampling along the plume axis from upgradient to downgradient should quantify electron donor/acceptor levels, determine any attenuation in contaminant levels, define the plume limits, and provide a picture of the background water chemistry. The right hand column of Table 8.3 provides a scoring system to make it easier to determine whether biodegradation is occurring.

Table 8.4 explains what the possible scores mean in terms of likelihood of natural attenuation ocurring. If the score exceeds 15 in the protocol of Wiedemeier et al. (1996) the next step (Step 2) is moved to. Tables 8.5 and 8.6 are, respectively, examples of sites which scored high and low for biodegradation of chlorinated organics. In the first case (Table 8.5), reducing conditions (low DO, low NO_3^-, high Fe^{2+}, low E_H,

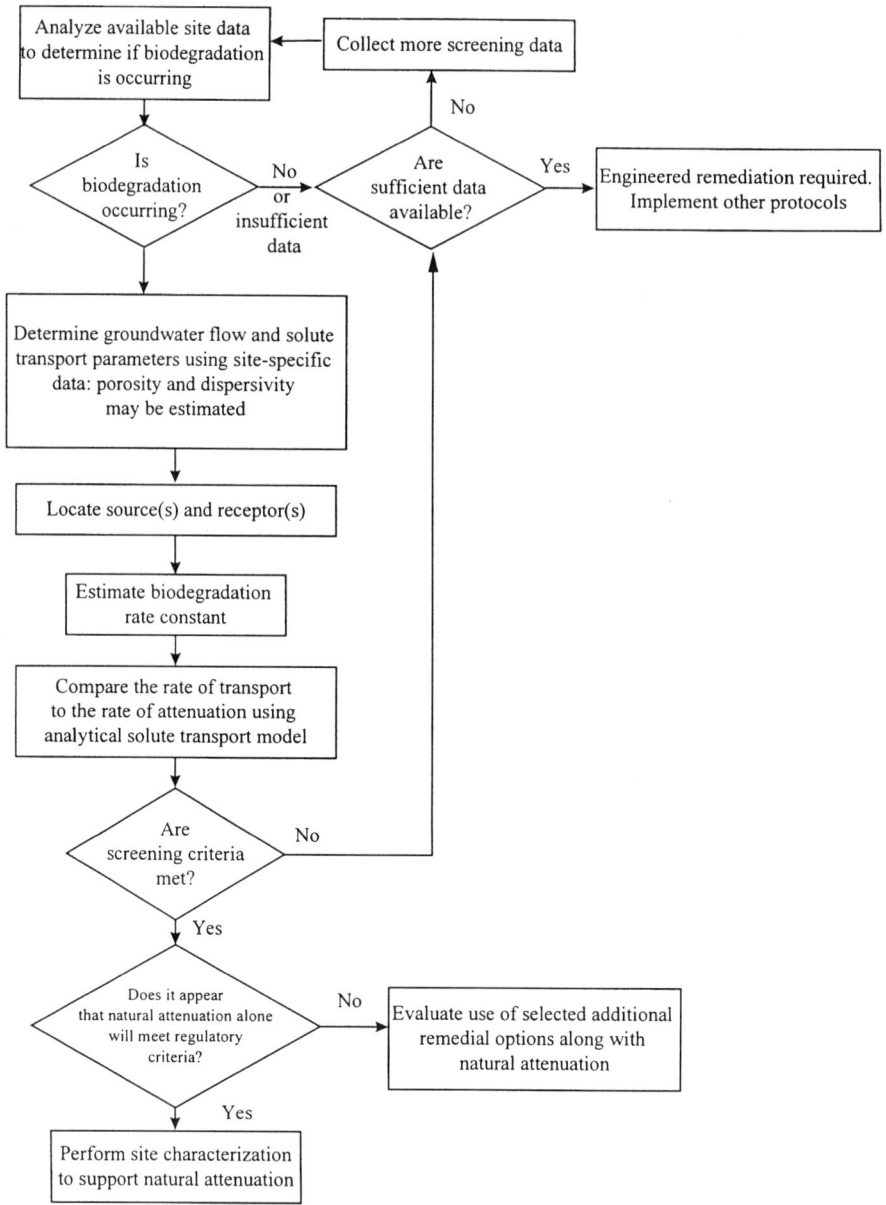

FIGURE 8.8 Screening procedure for chlorinated aliphatic natural attenuation protocol of Wiedemeier et al. (1996).

methane present), and the presence of abundant daughter products and chloride, in the most contaminated zone points strongly to degradation of the spilled PCE.

Table 8.6 describes a case where the prospects for natural attenuation are poor. Oxidizing conditions at the source (e.g., high DO, low Fe^{2+}, high sulfate, no methane,

TABLE 8.2
Soil and Groundwater Analytical Protocol

Analysis	Method/reference	Comments, data use	Frequency	Sample volume-container-preservation	Location
Matrix — Soil					
VOCS	SW8260A	Handbook method modified for field extraction of soil using methanol	Each soil sampling round	Collect 100 g of soil in a glass container with Teflon-lined cap; cool to 4°C	Fixed-base
TOC	SW9060, modified for soil samples	Procedure must be accurate over the range of 0.5 to 15% TOC	At initial sampling	Collect 100 g of soil in a glass container with Teflon-lined cap; cool to 4°C	Fixed
O_2, CO_2	Field soil gas analyzer		At initial sampling and respiration testing	Reusable 3-L Tedlar bags	Field
Fuel and Chlorinated VOCs	EPA Method TO-14		At initial sampling	1-L Summa cannister	Fixed-base
Matrix — Water					
VOCs	SW8260A	Handbook method; analysis may be extended to higher molecular-weight alkylbenzenes	Each sampling round	Collect water samples in a 40 mL VOC analysis vial; cool to 4°C; add hydrochloric acid to pH 2	Fixed-base
PAHs (optional)	Gas chromatography/Mass Spectroscopy method SW8310; High performance liquid chromatography method SW8310	Analysis only needed for regulatory complicance	As required by regulations	Collect 1 L of water in a glass container; cool to 4°C	Fixed-base

TABLE 8.2 (continued)
Soil and Groundwater Analytical Protocol

Analysis	Method/reference	Comments, data use	Frequency	Sample volume-container-preservation	Location
Oxygen	DO meter	Refer to method A4500 for a comparable laboratory procedure	Each sampling round	Measure DO on site using a flow-through cell	Field
Nitrate	Ion chromatography Method E300; anion method	Method E300 is a handbook method; also provides chloride data	Each sampling round	Collect up to 40 mL of water in a glass or plastic container; add H_2SO_4 to pH less than 2; cool to 4°C	Fixed-base
Iron(II)	Colorimetric HACH Method 8146	Filter if turbid	Each sampling round	Collect 100 mL of water in a glass container	Field
Sulfate	Ion Chromatography Method E300 or HACH Method 8051	Method E300 is a handbook method. HACH Method 8051 is a colorimetric method; use one or the other	Each sampling round	Collect up to 40 mL of water in a glass or plastic container; cool to 4°C	E300 = Fixed-base HACH Method 8051 = Field
Methane, ethane, or ethene	(Kampbell et al., 1989)	Method used by EPA researchers	Each sampling round	Collect water samples in 50 mL glass serum bottles with butyl gray/Teflon-lined caps; add H_2SO_4 to pH less than 2; cool to 4°C.	Fixed-base
Alkalinity	HACH alkalinity test kit model AL AP MG-L		Each sampling round	Collect 100 mL of water in a glass container	Field
E_H	A2580B		Each sampling round	Collect 100 to 250 mL of water in a glass container	Field

Analysis	Method	Reference	Frequency	Sample volume/container	Location
pH	Field probe with direct reading meter	Field	Each sampling round	Collect 100 to 250 mL of water in a glass or plastic container; analyze immediately	Field
Temperature	Field probe with direct reading meter	Field only	Each sampling round	Collect 100 to 250 mL of water in a plastic container	Field
Conductivity	E120.1/SW9050, direct reading meter	Protocols/Handbook methods	Each sampling round	Collect 250 mL of water in a glass container	Field
Chloride	Mercuric nitrate titration A4500 Cl-C	Ion chromatography Method E300; Method SW9050 may also be used	Each sampling round	Collect 250 mL of water in a glass container	Fixed-base
Chloride (optional; see data use)	HACH chloride test kit MODEL 8-P	Silver nitrate titration	Each sampling round	Collect 100 mL of water in a glass container	Field
Total organic carbon	SW9060	Laboratory	Each sampling round	Collect 100 mL of water in a glass container	Laboratory

Note: Analyses other than those listed may be required for regulatory compliance.

"SW" refers to the Test Methods for Evaluating Solid Waste, Physical, and Chemical Method (U.S. Environmental Protection Agency, 1986).

"E" refers to Methods for Chemical Analysis of Water and Wastes (U.S. Environmental Protection Agency, 1983).

"HACH" refers to the HACH company catalog.

"A" refers to Standard Methods for the Examination of Water and Wastewater (American Public Health Association, 1992).

"Handbook" refers to the AFCEE handbook to Support the Installation Restoration Program (IRP) Remedial Investigations and Feasibility Studies (RI/FS) (AFCEE, 1993).

"Protocols" refers to the AFCEE Environmental Chemistry Function Installation Restoration Program Analytical Protocols (AFCEE, 1992).

From Wiedemeier et al., 1996.

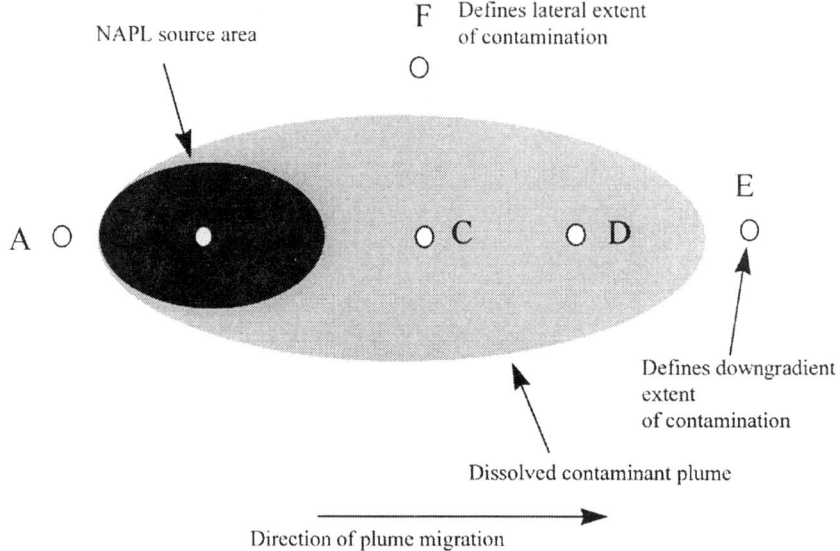

FIGURE 8.9 Sampling locations for chlorinated aliphatic natural attenuation protocol of Wiedemeier et al. (1996).

and a high E_H) guarantee that the contaminant (TCE) will not be reductively degraded at an appreciable rate. Confirmation of this is the lack of daughter products, and the fact that chloride levels are equivalent to background.

In the first case, Step 2 would be subsequently moved on to, and the screening procedure continued. For the latter case, another remedy besides natural attenuation must be considered. In Step 2 a number of standard fluid flow and transport parameters (gradient, hydraulic conductivity, porosity, dispersivity, contaminant K_d) are collected for further assessment of contaminant movement. Some of these can be estimated, but it is better to have site-specific values. Step 3 requires identification of nearest receptors. Step 4 involves the estimation of the biodegradation rate constant. Degradation of chlorinated solvents is generally assumed to be described by a first-order rate law. Consequently, rate constants can be calculated by the approaches shown in Figures 8.2 and 8.3. Alternatively, BIOSCREEN can be used to estimate a rate constant by trial-and-error fitting of measured values. Step 5 involves using a reaction-transport code to forward model plume migration towards receptors (For this, Wiedemeier et al. (1996) suggest using BIOSCREEN), and to perform a sensitivity analysis of the calculation. If contaminant levels are calculated to attenuate to levels desired by regulators, the screening procedure is continued to Step 6. If natural attenuation will not reach these levels, and there are no data gaps, other remediation methodologies must be considered. If a lack of relevant data prevents a clear assessment, more data must be acquired and Steps 1 through 5 must be repeated.

Step 6 requires that the satisfactory achievement of the screening criteria be scrutinized. Specifically, the consistency between predicted and observed plume movement must be assessed. The likelihood of the contaminant degrading before

Demonstrating Natural Attenuation

TABLE 8.3
Analytical Parameters and Weighting for Preliminary Screening

Analyte	Concentration in most contaminated zone	Interpretation	Points awarded
Oxygen[a]	<0.5 mg/l	Tolerated; suppresses reductive dechlorination at higher levels	3
Oxygen[a]	>1 mg/l	Vinyl chloride may be oxidized aerobically — but reductive dechlorination will not occur	-3
Nitrate[a]	<1 mg/l	May compete with reductive pathway at higher concentrations	2
Iron(II)[a]	>1 mg/l	Reductive pathway possible	3
Sulfate[a]	<20 mg/l	May compete with reductive pathway at higher concentrations	2
Sulfide[a]	>1 mg/l	Reductive pathway possible	3
Methane[a]	>0.1 mg/l	Ultimate reductive daughter product	2
	>1	Vinyl chloride accumulates	
	<1	Vinyl chloride oxidizes	3
E_H[a]	<50 mv against Ag/AgCl	Reductive pathway possible	<50 mv = 1 <-100 mV = 2
pH[a]	5 < pH < 9	Tolerated range for reductive pathway	
DOC	>20 mg/l	Carbon and energy source; drives dechlorination; can be natural or anthropogenic	2
Temperature[a]	>20°C	At >20°C biochemical process is accelerated	1
CO_2	>2× background	Ultimate oxidative daughter product	1
Alkalinity	>2× background	Results from interaction of carbon dioxide with aquifer minerals	1
Chloride[a]	>2× background	Daughter product of organic chlorine	2
H_2	>1 nm	Reductive pathway possible; vinyl chloride may accumulate	
	<1 nm	Vinyl chloride oxidized	3
Volatile fatty acids	>0.1 mg/l	Intermediates resulting from biodegradation of aromatic compounds: carbon and energy source	2
BTEX[a]	>0.1 mg/l	Carbon and energy source; drives dechlorination	2
PCE[a]		Material released	
TCE[a]		Material released or daughter product of PCE	2[b]

TABLE 8.3 (continued)
Analytical Parameters and Weighting for Preliminary Screening

Analyte	Concentration in most contaminated zone	Interpretation	Points awarded
DCE[a]		Material released or daughter product of TCE; if amount of *cis*-1,2-dichloroethene is greater than 80% of total dichloroethene, it is likely a daughter product of TCE	2[b]
Vinyl chloride[a]		Material released or daughter product of dichloroethenes	2[b]
Ethene/ethane[a]	<0.1 mg/l	Daughter product of vinyl chloride/ethene	>0.01 mg/l = 2 >0.1 = 3
Chloroethane[a]		Daughter product of vinyl chloride under reducing condition	2
1,1,1-Trichloroethane[a]		Material released	
1,1-dichloroethene		Daughter product of TCE or chemical reaction of 1,1,1-trichloroethane	

[a] Required analysis.
[b] Points awarded only if it can be shown that the compound is a daughter product (i.e., not a constituent of the source NAPL).

TABLE 8.4
Interpretation of Points Awarded During Screening

Score	Interpretation
0 to 5	Inadequate evidence for biodegradation of chlorinated organics
6 to 14	Limited evidence for biodegradation of chlorinated organics
15 to 20	Adequate evidence for biodegradation of chlorinated organics
>20	Strong evidence for biodegradation of chlorinated organics

TABLE 8.5
Strong Evidence for Biodegradation of Chlorinated Organics

Analyte	Concentration in most contaminated zone	Points awarded
DO	0.1 mg/l	3
Nitrate	0.3 mg/l	2
Iron(II)	10 mg/l	3
Sulfate	2 mg/l	2
Methane	5 mg/l	3
E_H	–190 mv	2
Chloride	3x background	2
PCE (released)	1000 µg/l	0
TCE (none released)	1200 ∝γ/l	2
cis-1,2-DCE (none released)	500 µg/l	2
Vinyl chloride (none released)	50 µg/l	2
	Total Points	23

TABLE 8.6
Poor Evidence for Biodegradation of Chlorinated Organics

Analyte	Concentration in most contaminated zone	Points awarded
DO	3 mg/l	–3
Nitrate	0.3 mg/l	2
Iron(II)	Not detected	0
Sulfate	10 mg/l	2
Methane	Not detected	0
E_H	100 mv	0
Chloride	Background	0
TCE (released)	1200 ∝γ/l	0
cis-1,2-DCE (none released)	Not detected	0
Vinyl chloride	Not detected	0
	Total Points	1

affecting receptors must be ascertained as well. If the first is evident, and the latter appears promising as well, the screening procedure is completed, and the additional site characterization step in Figure 8.7 is moved to. The steps which follow are pretty straight-forward, and are quite similar to the analogous steps outlined in the AFCEE protocol for natural attenuation of fuel hydrocarbons.

An estimated 20% of the Air Force sites contaminated with chlorinated solvents might prove treatable by natural attenuation (Wiedemeier et al., 1996). In general, the Air Force sites have longer distances between contaminant source and nearest receptors than non-Air Force sites. At the same time institutional control might be easier maintained at the Air Force sites. Both would favor natural attenuation as the treatment approach. Consequently, the Air Force might have a higher fraction of sites treatable by natural attenuation than the private sector (E. Glaza, personal communication). Note though that this may be compensated somewhat if the non-Air Force sites are, taken individually, smaller in volume.

It should be re-emphasized that the demonstration of natural attenuation as outlined above is not cheap. The costs of analytical efforts certainly rival the costs of conventional remediation efforts. Nevertheless, it is reasonable to expect that the accumulation of institutional experience with remediation by natural attenuation will ultimately allow the demonstration and long-term monitoring requirements to be sharply reduced in time and money.

METALS

To our knowledge, there is no protocol for demonstrating natural attenuation of metals. Nor can the protocols for natural attenuation of organic contaminants be used without modification as templates. Natural attenuation of organic contaminants is generally demonstrated using a wealth of evidence argument, and as many independent lines of evidence pointing to reductions in contaminant mass as can be had are brought together to convince a regulatory agency of a particular degradation pathway. The four most effective components to such an argument are (1) evidence of contaminant loss in the field, (2) variations in electron donor/acceptor levels, (3) appearance of degradation by-products, and (4) soil microcosm studies done in the lab. Making the case for natural attenuation by irreversible sorption is harder for a number of reasons. The appearance of by-products, or variation in acceptor/donor levels, cannot be used to monitor irreversible sorption primarily because, when a contaminant, such as lead, sorbs it will probably displace some other cation such as Ca^{2+}, which is likely to be far more abundant in solution. When Cs^+ sorbs to a clay, chances are that it will be present in only trace amounts, and far less abundant in solution than the Na^+ or K^+ it displaces. As a result, while irreversible sorption of trace contaminants will dramatically affect solution levels of the latter, changes in other background metal concentrations will more than likely be minimal. Consequently, the latter probably can't be used to demonstrate irreversible sorption.

Metal desorption tests in the laboratory are best used to bracket mechanisms. Desorption rates measured over weeks to months in the lab don't say a whole lot about reactions occurring in soils over time-spans of decades or more. There is the chance that measuring desorption rates at higher temperatures, where rates are faster,

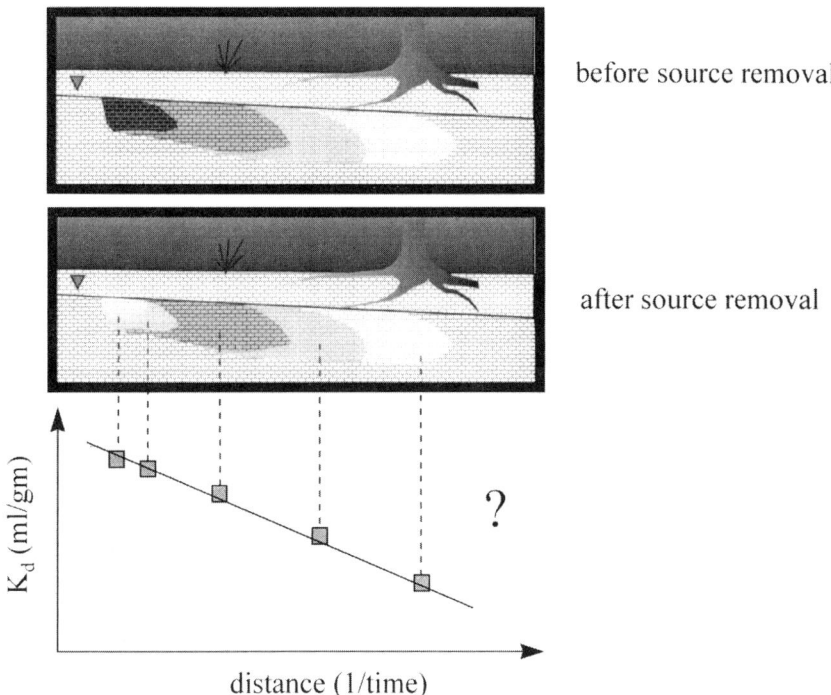

FIGURE 8.10 Variation of K_d over time and distance down a plume axis. See text.

might give some useful long-term predictions. Alternatively, molecular models which simulate the chemical reactivity of metal-mineral pairs might be used to forward model desorption rates which can't be measured over laboratory time scales. Before doing so, though, the computational results should be rigorously calibrated against short-term desorption measurements.

Lab experimental results and molecular model calculations are only a poor substitute for field evidence demonstrating irreversible sorption. To our knowledge there is no proven way in which the progressive uptake of contaminants can be demonstrated in the field, though there is no shortage of experimental evidence which says this often occurs. One potential approach to demonstrating irreversible sorption might take advantage of the general observation the K_d's appear to increase the longer a metal is in contact with a mineral. A hypothetical case of a metal-rich plume is shown in Figure 8.10. Once the metal source is removed, a progressive freshening of the adjacent solutions will occur due to recharge. Note that metals farthest from the source of the plume will have been in contact for the shortest period of time, whereas the metals nearest the source will have been in contact with the soil for the longest period of time. If irreversible sorption of the metals has occurred, the K_d should increase with time and be greater at the source of the plume, and smallest at the end of the plume. As a litmus test of irreversible sorption, K_d's might be measured in the field by taking samples along the axis of the plume. Specifically, at a series of points along the axis of plume movement soil + solution samples might be taken and dissolved metal levels be analyzed in the latter. A digestion, or corrosive leach, of the mineral surface, followed by analysis of the supernatant, would allow the sorbed/occluded fraction to be determined. Combination of these two analyses would then provide a field "K_d". Doing the same thing closer and closer to the contaminant source would in turn provide a trend in K_d with source proximity, as well as with metal-mineral contact time. If K_d's increase with time, as shown in Figure 8.10, one might plausibly argue that irreversible sorption was occurring (a flat line would indicate no irreversible sorption). Obviously, this approach lumps the Δ fraction in with the exchangeable fraction, so the K_d is actually monitoring the sum of 2 reactions: sorption and sequestering. The increase in such an operative K_d comes about solely due to the sequestering step. The steepness of the K_d-distance curve might also be used to predict how long contaminants were likely to remain in solution; i.e., the time required for complete natural attenuation of the site contamination.

At this point the technique is simply a hypothesis. Nevertheless, there are a number of field observations pointing to irreversible sorption control over contaminant transport. Alexander's results demonstrating diminished bioavailability of organic molecules at mineral surfaces over and above any biodegradation supports the notion of K_d increase with time. Similarly, Brannon et al. (1995) observed an increase in PCB and PAH binding to sediment samples with time.

Some of the irreversible sorption of organic contaminants involves their sequestration in dead-end pores. Dead-end pores can be imaged by electron microscopy (e.g., Berner et al., 1980; Dorn and Brady, 1995) (see Figure 8.11), and, if the contaminant is a metal, their contents occasionally determined (see Figure 8.12). It appears that the fraction of contaminants in dead-end pores is less bioavailable (Alexander, 1995) relative to contaminants sorbed on the outer edges of minerals in close contact with moving fluids. Nevertheless, some organic contaminants aren't biodegraded or incorporated into intraparticulate porosity. Alexander (1995) examined field records which indicated that DDT, aldrin, edieldrin, heptachlor, chlordane, kepone, nonylphenol and a linear alkylbenzene solfonate were sequestered. Lab and field studies cited by Alexander (1995) pointed to the same destination for 1,2-dibromomethane, the herbacide simazine, napthalene, and phenanthrene. Nevertheless, 2,3,7,8-TCDD, PCBs, and polybrominated biphenyls, while strongly sorbed, don't lose toxicity as might be predicted if they were irreversibly sorbed (Alexander, 1995). Presumably the process is the sum of a mineralogic and a contaminant-specific component. The availability of mineral micropores, the distribution of hydrophobic to hydrophilic sites (Alexander, 1995), and the geometric make-up of the pores probably control the former. The contaminant side of the equation probably depends on the relative affinity of the various functional groups for the mineral surfaces. Until this breakdown is proven it is critical that the implied non-leachability of contaminants be demonstrated on a site-by-site basis. This might be done by way of a TCLP, or other standard (non-corrosive) leaching test. The latter point is important. Acetate and citrate buffers used in, respectively, the TCLP

Demonstrating Natural Attenuation

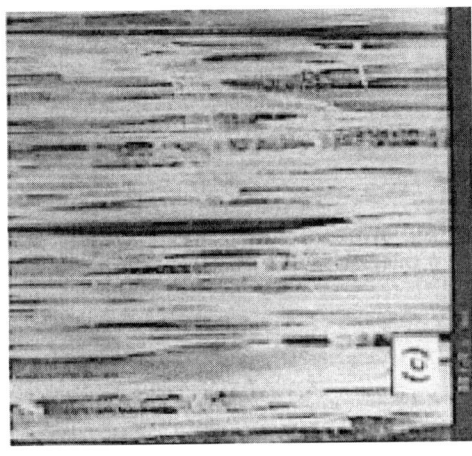
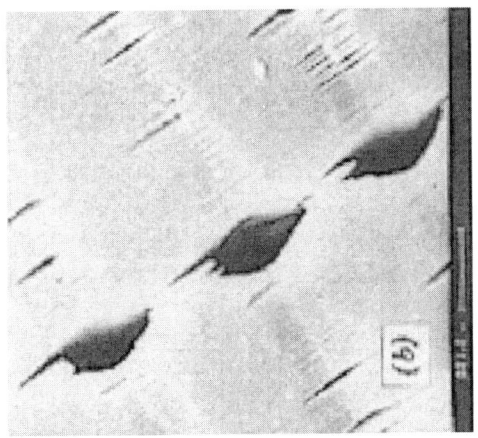
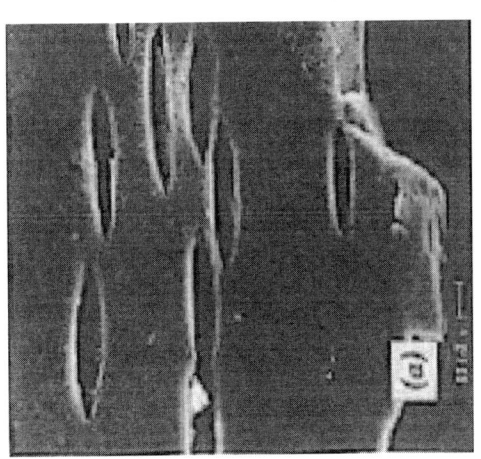

FIGURE 8.11 Pits on surfaces on hypersthene (a) and diopside (b) and (c). (From Berner, R. A. et al., *Science*, 207, 1205, 1980. Copyright 1980 American Association for the Advancement of Science. With permission.)

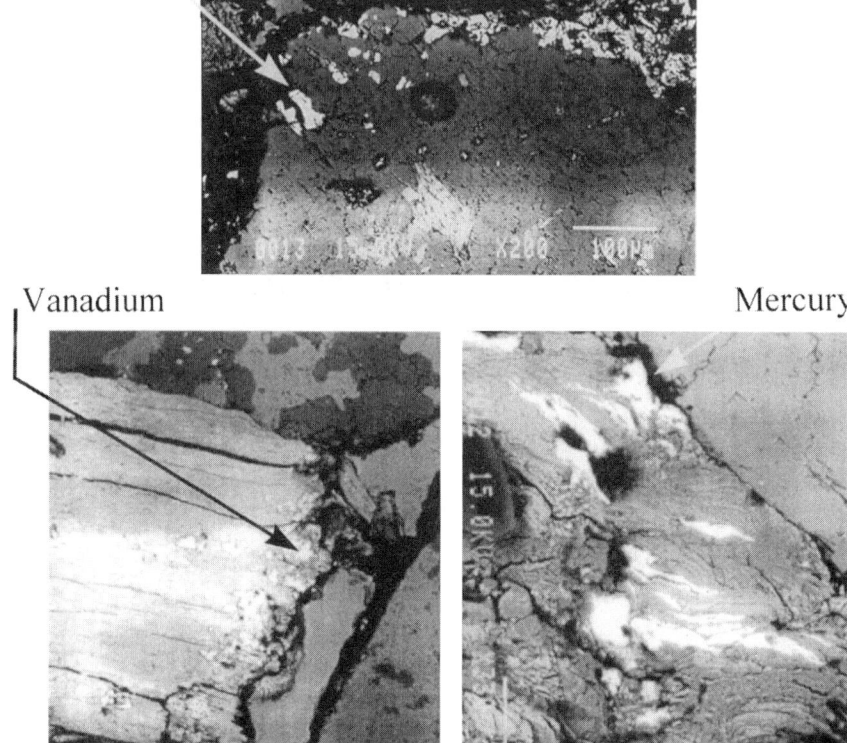

FIGURE 8.12 Top: orthoclase feldspar grain (lower half of the image) hosts a coating of cadmium-rich particles (probe analyses indicate a cadmium concentration of ~30% in bright particles). The cadmium-rich particles are all connected by a mass of algal filaments (that appear dark in back-scattered electron image). In addition, the cadmium appears to be migrating into the grain and precipitating within the pore spaces. The scale bar on the bottom is 100 microns. Bottom: granitic pebble hosts mercury and vanadium (as determined by microprobe analyses). The contrast on the BSE was turned as low as possible in order to prevent the brightness of the mercury from "swamping" the rest of the image. Hence, all of the surrounding silicate minerals have about the same brightness (except the organics that appear dark gray). The mineral in the upper left is quartz, and the material that hosts the mercury is weathered biotite. Both minerals were collected from the San Pedro River near Benson, AZ. (Courtesy of Prof. R. I. Dorn — Arizona State University).

and WET, both will dissolve calcium carbonate, and preferentially mobilize any metals sorbed onto, or occluded in, a calcium carbonate mineral. A citrate buffer will also destabilize iron hydroxides, and mobilize any sorbed metals. Note that citrate forms strong ion pairs with a number of contaminant metals and radionuclides. The net result is that leachability will tend to be exaggerated. In effect, using a leach solution which dissolves the sequestering phases will obscure natural attenuation of sorbed contaminants. For this reason, it is more useful to perform leach tests using solutions similar in composition to those in contact with the soil or aquifer matrix.

Sorbed radionuclides are generally present in trace quantities. Due to this low concentration, and also to their radiotoxicity, surface analytical and leach tests are neither cheap nor easy. Millions of dollars have been spent by DOE, its predecessors, and foreign agencies to measure K_d's and identify the primary controls on radionuclide sorption and desorption. Transport models using radionuclide K_d's are therefore (arguably) better constrained than for most contaminants. At the same time, there is extensive institutional experience in DOE and the NRC in applying these models to determine the potential for natural attenuation of radionuclides.

PROTOCOL FOR DEMONSTRATING NATURAL ATTENUATION OF METALS

Condensing the preceding sections, and following the overall structure of Wiedemeier et al. (1996), a hypothetical protocol for demonstrating natural attenuation of contaminant metal might entail:

1. Review available site data
2. Develop preliminary conceptual model and assess the potential for natural attenuation
3. If needed, perform additional site characterization to support natural attenuation
4. Update conceptual model
5. Simulate long-term site behavior
6. Perform an exposure pathways analysis
7. If natural attenuation is acceptable, prepare long-term monitoring plan
8. Present results to regulators

These will be dealt with in turn.

Step 1. Review available site data. As noted above this should involve an assessment of the contaminant source, and should also provide a useable hydrologic model, roughly locating receptors and pathways.

Step 2. Develop preliminary conceptual model for natural attenuation. This will depend on the metals (and other inorganics) and the soil. Table 8.9 lists metals and likely reasonable natural attenuation pathways. The latter were taken largely from the analysis in the preceding sections. When sorption is mentioned, it is assumed that irreversible sorption is potentially important as well.

Table 8.7 also points up the types of data needed to demonstrate natural attenuation of metals. Data needs depend primarily on whether the likely fate of the compound is as a component of an insoluble solid, a sorbed contaminant, or, possibly, a species occluded on an iron hydroxide or carbonate mineral surface, or irreversibly sorbed to an innerlayer clay site. Constructing a preliminary model for natural attenuation revolves around determining the likelihood of the pathway being important at the particular site. Step 3, performing additional site characterization, will, in many cases, involve the gathering of additional soil and groundwater chemical data to answer this question. A metal-by-metal listing of relevant data needs is shown in Table 8.8.

Once the relevant data has been obtained, standard geochemical codes (see Chapter 5) can be used to calculate whether thermodynamics favors contaminant levels to be limited by the formation of an insoluble phase (e.g., Ba^{2+} by $BaSO_4$ growth). Geochemical modeling to support uptake by sorption is not far enough along to be as a stand-alone demonstration of metal sorption. Instead, uptake by sorption can be demonstrated by: (1) showing that the sorbing phase (e.g., Fe-hydroxides, calcium carbonate) is present in soils and is thermodynamically stable through a solubility calculation; and (2) showing that an appreciable fraction of the compound is associated with that phase. The latter is most directly done through sequential soil leaching procedures which dissolve specific minerals, along with any sorbed material. Citrate-Dithionate solutions remove iron hydroxides from soils. HF removes silicates. H_2O_2 removes organic matter. As noted earlier, acid acetate buffer solutions remove calcium carbonate.

The sequential extraction of mineral and organic soil components is described in detail by Jackson (1969), in Dragun (1988), and in Yong et al. (1993). The latter approach relies on a combination of equilibrium speciation calculations and sequential leaching procedures. As noted earlier, chemical speciation is critical to the prediction of metal transport and toxicity. Sequential extraction analysis is particularly useful because it provides a rough measure of the capacity for a given soil (or backfill) to attenuate heavy metal toxicity (Yong et al., 1993).

In the approach of Yong et al. (1993) metals are assumed to populate one of 5 distinct pools; exchange sites, in/on carbonate minerals, in/on metal (hydr)oxides, in/on organic matter, and everything else. Metals associated with "everything else" are considered to be those heavy metals which have managed to work themselves tightly into silicate mineral matrices. Heavy metals on exchange sites are loosely held by electrostatic attraction to clay minerals, organic matter, and amorphous materials (Yong et al., 1993). Washing of contaminated soils in high levels ($1M$) of dissolved salts such as $MgCl_2$, $NaNO_3$, and $CaCl_2$ removes most exchangeable heavy metals from the solid phase, and into solution. Analysis of the supernatant for the metal(s) of interest then quantifies the fraction on exchange sites. At the same time, such leaches do not otherwise appreciably disturb heavy metals in the other pools. This is important because the idea is to interrogate the soil with progressively more corrosive leaches.

Heavy metals associated with carbonate minerals are removed from soils by exposing the latter to acid solutions which destroy the carbonate host. A 1M HOAc-NaOAc solution (Ac = acetate) is recommended to remove calcite and dolomite, two of the

TABLE 8.7
Natural Attenuation Pathways for Metals (and Other Inorganics)

Chemical	Natural attenuation pathways	Caveats, special data needs
Pb^{2+}	Sorption to iron hydroxides, organic matter, carbonate minerals, formation of insoluble sulfides.	Low pH destabilizes carbonates, iron hydroxides. Comingled organic acids and chelates (e.g., EDTA) may decrease sorption. Low E_H dissolves iron hydroxides, but favors sulfide formation.
CrO_4^{2-}	Reduction by organic matter, sorption to iron hydroxides, formation of $BaCrO_4$	Low pH destabilizes carbonates, iron hydroxides. Low E_H dissolves iron hydroxides. Are reductants available?
As(III or V)	Sorption to iron hydroxides, formation of sulfides	Low pH destabilizes carbonates, iron hydroxides. Low E_H dissolves iron hydroxides.
Zn^{2+}	Sorption to iron hydroxides, carbonate minerals, formation of sulfides	Low pH destabilizes carbonates, iron hydroxides. Comingled organic acids and chelates may decrease sorption. Low E_H dissolves iron hydroxides.
Cd^{2+}	Sorption to iron hydroxides, carbonate minerals, formation of insoluble sulfides.	Low pH destabilizes carbonates, iron hydroxides. Comingled organic acids and chelates may decrease sorption. Low E_H dissolves iron hydroxides, but favors formation of sulfides.
Ba^{2+}	Sorption to iron hydroxides, formation of insoluble sulfate minerals	Low pH destabilizes carbonates, iron hydroxides. Low E_H dissolves iron hydroxides. What are sulfate levels?
Ni^{2+}	Sorption to iron hydroxides, carbonate minerals	Low pH destabilizes carbonates, iron hydroxides. Comingled organic acids and chelates may decrease sorption. Low E_H dissolves iron hydroxides, but favors sulfide formation.
Hg^{2+}	Formation of insoluble sulfides	Methylated by organisms
NO_3^-	Reduction by biologic processes	

RADIOACTIVES

UO_2^{+2}	Sorption to iron hydroxides, precipitation of insoluble minerals, reduction to insoluble valence states	Low pH destabilizes carbonates, iron hydroxides. Comingled organic acids and chelates may decrease sorption. High pH and/or carbonate levels decrease sorption. Low E_H dissolves iron hydroxides.
Pu(V and VI)	Sorption to iron hydroxides, formation of insoluble hydroxides	May move as a colloid. Low E_H dissolves iron hydroxides.
Sr^{2+}	Sorption to carbonate minerals, formation of insoluble sulfates	Low pH destabilizes carbonates.
Am^{3+}	Sorption to carbonate minerals	Low pH destabilizes carbonates. High pH increases solubility of Am-carbonate minerals.

TABLE 8.7 (continued)
Natural Attenuation Pathways for Metals (and Other Inorganics)

Chemical	Natural attenuation pathways	Caveats, special data needs
Cs^+	Sorption to clay innerlayers	High NH_4^+ levels may lessen sorption. How abundant are clays?
I^-	Sorption to sulfides, organic matter	Sorbs to very little else.
TcO_4^-	Possible reductive sorption to reduced minerals (e.g., magnetite), forms insoluble reduced oxides and sulfides.	Sorbs to very little else.
Th^{4+}	Sorption to most minerals, formation of insoluble hydroxide	May move as a colloid
Co^{2+}	Sorption to iron hydroxides, carbonate minerals	Low pH destabilizes carbonates.

TABLE 8.8
Data Needs for Natural Attenuation of Metals

Chemical	Data needs
Pb^{2+}	Iron hydroxide availability; pH, alkalinity, and Ca^{2+} levels to answer if calcium carbonate is stable. E_H, and if E_H is low, sulfide levels. Organic carbon content.
CrO_4^{2-}	E_H, electron donor levels, pH (reduction rates are faster at low pH). See chromate example in Chapter 7.
As(III or V)	E_H and if E_H is low, sulfide levels.
Zn^{2+}	Iron hydroxide availability; pH, alkalinity, and Ca^{2+} levels to answer if calcium carbonate is stable. E_H, and if E_H is low, sulfide levels.
Cd^{2+}	Iron hydroxide availability; pH, alkalinity, and Ca^{2+} levels to answer if calcium carbonate is stable. E_H, and if E_H is low, sulfide levels.
Ba^{2+}	Sulfate levels.
Ni^{2+}	Iron hydroxide availability; pH, alkalinity, and Ca^{2+} levels to answer if calcium carbonate is stable. E_H, and if E_H is low, sulfide levels.
Hg^{2+}	E_H, and if E_H is low, sulfide levels.
UO_2^{+2}	Iron hydroxide availability, pH, availability of reducing compound
Pu(V and VI)	Iron hydroxide availability, pH, availability of reducing compound
Sr^{2+}	Iron hydroxide availability; pH, alkalinity, and Ca^{2+} levels to answer if calcium carbonate is stable.
Am^{3+}	Iron hydroxide availability; pH, alkalinity, and Ca^{2+} levels to answer if calcium carbonate is stable.
Cs^+	Clay content, cation exchange capacity.
I^-	Metal sulfide mineral content
TcO_4^-	E_H, and if E_H is low, sulfide levels.
Co^{2+}	Iron hydroxide availability; pH, alkalinity, and Ca^{2+} levels to answer if calcium carbonate is stable.

most common carbonate phases, while leaving behind metal (hydr)oxides and organic matter. Extraction of the metal (hydr)oxide fellow travelers must not, in turn, cause the release of heavy metals from organic matter or the tightly held silicate fraction. Yong et al. (1993) recommend a combination of $1M$ hydroxylamine hydrochloride cut with one part in four by volume of acetic acid.

Ascribing heavy metals to soil organic matter is difficult because metals associated with organic matter is the sum of both exchangeable and organic-specific sites. The first will be lumped in with the first sequential extraction. The latter are generally removed by oxidation of the organic matter itself. The "everything else"-silicate fraction is not considered by Yong et al. (1993) to be large. In any case, digestion in HF, and subsequent analysis, can put a number on the latter.

If the primary natural attenuation pathway is irreversible sorption, *Step 5, the long-term modeling of site behavior,* is arguably more difficult for metals-contaminated sites than it is for sites contaminated with fuel hydrocarbons. As noted in the previous section, the kinetics of irreversible metal sorption is much less clear than is the measurement and application of biodegradation rate constants. For compounds which form insoluble minerals, one can calculate the solubility-based compound level in solution at the source, at which point any number of transport codes can be used to estimate the amount of dilution likely to occur downgradient (Just because a compound is calculated to form a relatively insoluble phase does not mean the latter can be assumed to automatically be present — it must be identified at the site as well). The sequential leaching procedures alluded to in the preceding paragraph lend themselves to the measurement of an "effective" K_d if the dissolved level of the metal is measured as well, and if sorption is the primary control over dissolved concentrations. A K_d can likewise be used as input in transport codes to estimate contaminant levels over distance. For both the solubility and sorption-limited cases, some effort must be made to determine if the conditions limiting compound movement at present are likely to hold over the long term.

Step 6, the performance of an exposure pathways analysis, is only different for metals to the extent that ingestion of the solids may be important. *Step 7, the preparation of a monitoring plan,* requires the monitoring of those soil parameters which limit transport (E_H, pH, availability of sorbents, etc.), and is fairly straightforward.

BACKGROUND READING

Dragun J. (1988) The Soil Chemistry of Hazardous Materials — Hazardous Materials Control ResearchInstitute, Silver Spring, MD.

9 The Present

The assumption that every industrial by-product that ends up in soils or natural waters will become a "time bomb" having inevitable repercussions for human health is wrong. The evidence outlined above points to a very substantial self-cleansing capacity of soils. The combined, and sometimes individual, effects of adsorption, precipitation, biodegradation, dilution and dispersion often render large fractions of otherwise dangerous compounds inert or harmless, often at a rate comparable to much more expensive engineered solutions. Consequently, this argues for a different approach to remediation of environmental contamination; one that takes advantage of natural attenuation processes, or at the very least, does not assume that soils are direct conduits connecting waste sites to drinking water. As a confirmatory aside, the past few months have witnessed the marketing of a number of "new" remediation tools/approaches including; phytoremediation (plants that eat contaminants), special humics that sequester contaminants, engineered organisms, mineral-based backfills that isolate wastes for long-periods of time, etc. Note that a mixture of minerals, humics, microorganisms, and plants, is, well, a soil.

CERCLA is that unique program that virtually *every* sector of the political spectrum agrees must be changed. But how to change it? The quick answer is "change the liability scheme." We do not believe that there is a principled way to reform CERCLA's liability scheme (This is not to imply that we believe the present scheme to be particularly fair — but fairness was never the point. CERCLA's main goal was to amass sufficient funds to perform cleanups. Even if only a fraction of the fund goes to actual remediation, that amount is much greater than would otherwise have existed). Even if CERCLA's liability scheme were revised to delete retroactive liability, the most popular proposed quick-fix, the problem still wouldn't go away. As long as the "cleanup" continues to be an open-ended obligation that runs into the tens of millions of dollars, parties will continue to pay their attorneys first. If "no-retroactive liability" were achieved, the remaining private parties would fight all the harder. And if they got out, or refused to participate or went bankrupt, who would be left? The government?

In some ways this would be fitting because Congress created the problem when it stated its preference for active treatment at NPL sites (CERCLA Section 121). EPA, in the NCP, compounded the problem by excluding serious consideration of cleanup technologies that *did not* employ "big engineering" projects. CERCLA didn't just create a cottage industry for environmental lawyers, it created a huge environmental consultant/engineer industry to build and run the big engineering projects along with a regulatory apparatus to oversee the consultants and make certain that the big engineering projects somehow tried to comply with the near-impossible mandate contained in CERCLA.

Ultimately, liability reform or no, Congress must inject *some* reality into the cleanup methodology EPA employs. That is not to say that Congress should simply give up on solving the problem, though that is the only alternative to meaningful reform. Meaningful reform would require consideration of natural attenuation processes first, followed by big engineering projects for those sites where natural attenuation processes are unavailing, and where it can be demonstrated that the extra effort is actually reducing health risk. Though the current methodology *permits* the consideration of natural attenuation in cleanup selection, it does not *require* it, and the NCP, with its emphasis on big engineering exercises, discourages it.

The primary purpose of environmental cleanups should be to minimize health risk in a timely and cost-efficient fashion. Under the approach being followed today, cleanup is hardly timely. Only a tiny fraction of the 330,000 sites pointed out in 1993 by the EPA have been cleaned up. But what of health risk minimization? Exposure to Superfund sites apparently does impair health, at least somewhat. The documented effects include birth defects caused by exposure to volatile organic contaminants, as well as speech and hearing impairments caused by exposure to TCE in early childhood, or in the womb (Lybarger and Spengler, 1996). In addition, there is decreased pulmonary function observed in persons exposed to emissions from smelters and phosphate ore-processing facilities. Nevertheless, the Agency for Toxic Substance and Disease Registry (ATSDR) states that "in aggregate, proximity to hazardous waste sites seems to be associated with a small to moderate increased risk", and there seems to be no clear connection between site exposure and cancer (Lybarger and Spengler, 1996).

The health risks being reduced (assuming for the sake of argument that remediation is actually occurring) are, by some estimates, relatively minor, yet the projected outlay is stunning. If the object of the cleanup is to decrease risk to ourselves and the environment, we must make the linkage between risk and action clear. None can argue with the statement that the sites posing the worst and most immediate health risk should be cleaned up first. This is not what is happening now though. In fact, the most contaminated federal hazardous waste sites haven't even been evaluated for inclusion on the NPL (GAO, 1996). Before reviewing the various approaches which have been proposed for balancing risk and remediation, it is important to give some background as to how risk is calculated and how sites are prioritized.

RISK

Risk to human health is a concept that is not easily quantified and standardized. In practice, it is difficult to measure a group value of what is "acceptable" risk (e.g., safe maximum driving speeds or secondary contact to cigarette smoke). Congress has not defined "acceptable" risk; except where a zero risk approach is mandated. Generally, Congress delegates responsibility to regulatory agencies such as EPA (or the state or tribal equivalents) to set standards and requirements that address acceptable risk from contaminants in the environment. Any predictive tool for long-term performance can only address the compliance with the regulatory or performance standard as set out by regulation and cannot address "what is safe" or "what is acceptable risk" as developed by public consensus or political action.

In the Superfund program a clear distinction is made between risks to human health and risks to the environment. Risk involves both toxicity and exposure, and risk assessment is the process by which the total risk to human health and the environment is calculated if the site is not cleaned up. Risk assessment involves: (1) data collection and evaluation, (2) toxicity assessment, (3) exposure assessment, and (4) risk characterization. Toxicity is further split into cancer and non-cancer effects. Exposure assessment looks at current and future exposures. Risk characterization seeks to predict the net effect of toxicity and exposure. Cancer risk is approximated by quantifying the additional risk of cancer incidence involved in exposure to contaminants. Note that this added risk is calculated from the ambient cancer risk of 25 to 33% (0.25 to 0.33). The latter rate varies geographically, from age group to age group, and depends on a variety of dietary and health factors. Non-cancer risk is calculated differently, as the extent to which exposure exceeds a limit defined by the EPA. The latter limit, the Acceptable Daily Intake (ADI) or Reference Dose (RfD), is thought to represent the amount of exposure to a given contaminant that humans can receive on a daily basis without adverse health effects. The Hazard Quotient (HQ) for a single contaminant is defined as: HQ = Predicted Exposure/ADI. The overall Hazard Index (HI) for a site containing a number of contaminants is the sum of all of the individual HQ's, and is assumed to represent the total non-cancer risk of a site. Note that risk from exposure to carcinogenic contaminants is assumed to be cumulative, whereas the dose per unit time is important for non-carcinogenic contaminants. While reductions in contaminant levels by natural processes lessen the impact of both carcinogenic and non-carcinogenic contaminants, the net decrease in risk is calculated differently (see Borgert et al., 1995). Environmental risks are less clear. Environmental impacts might include habitat loss, species elimination or decrease in numbers, loss or reduction of use of a resource (e.g., Greenberg and Anderson, 1984). In the absence of a numerical methodology to calculate risk, "assumptions and value judgments play a key and controversial role in ecological risk issues" (Milloy, 1995).

Milloy (1995) has shown that human risk (cancer and non-cancer) are almost always overestimated, sometimes by several orders of magnitude. Although not the specific target of the present work, it is useful to briefly tabulate the assumptions which give rise to this perceived overestimation in the toxicity assessment (Milloy, 1995). These assumptions are that:

1. Substances which cause animal cancers cause human cancers.
2. Noncancerous tumors found in lab animals are cancerous.
3. Cancer risk scales linearly from high levels used in the lab to low levels present in the field.
4. Human susceptibility to cancer is identical to that of the most susceptible lab animal.
5. There is no safe exposure to any carcinogen.
6. Each member of a chemical class is as carcinogenic as the most carcinogenic member of that class.
7. All cancers are equivalent (e.g., skin cancer, which is easily treatable, is equivalent to lung cancer, which is not).
8. Cancer latency and life span are irrelevant to cancer risk.

And for non-carcinogenic contaminants:

1. Biologic effects observed in the lab are assumed to be adverse health effects.
2. Humans are 100 times more susceptible than are lab animals.
3. Hazard quotients can be summed.

Risks are routinely overestimated by several orders of magnitude (Milloy, 1995), because of the desire to be conservative. Because real risk is buried beneath layers of hypothetical risk it is very difficult to identify the really bad sites which require remediation. They all look bad.

PRIORITIZATION

Risks should be evaluated realistically and sites should be prioritized *before* remediation (e.g., NRC, 1990). Nevertheless, existing risk assessment and prioritization methods do not completely and consistently identify Superfund facilities and sites that present the greatest risks to the public health and environment (GAO, 1996). In part this is because:

1. There are inadequate data (e.g., waste inventories) to support site assessment. At the same time some parameters (e.g., K_h and K_d) are site-specific.
2. The assessment process doesn't reflect all of the significant types of contaminants and pathways.
3. Legal requirements, such as multiparty agreements in the case of DOE, often direct prioritization but do not reflect a significant reduction of the greatest nationwide risks.

For example, within EPA, DOE, and DoD, risk assessment and prioritization is *quantitatively* scored, yet the scoring in the individual categories are actually *qualitatively* based on expert judgment (NAS/NRC, 1995). This means that the methods can't account for a lack of data or an incomplete understanding of the interaction between the contaminant and the groundwater/soil matrix. One way around this is through the application of probabilistic decision analysis (see e.g., Freeze et al., 1990; Miller et al., 1993; Rautman, 1996; Webb et al., 1993), which uses statistical methods to compensate for knowledge or data gaps. Using the case of site characterization, the approach might assign actual cash costs to contaminant analyses and balance these outlays against a potential cost of liability if analyses are not done and residual contamination is detected later. Probabilistic decision analysis is useful because remediation costs and risk reduction are constantly being balanced against each other. Another important tool is the total system performance approach which seeks to iteratively model the dynamic interaction between the contaminant and the groundwater/soil matrix and the combined response to any engineered remediation efforts (Davis et al., 1996). We use similar approaches, for example, when buying a car, selecting an area to live in, or choosing a career. We follow the analogy between buying a car and cleaning up a site and show that many of the decision points are quite similar. Deciding that a new car is needed is analogous to the site characterization phase. Getting more data on the various makes and models is analogous to weighing the various benefits and costs of remediation strategies (e.g., pump-and-treat vs. soil venting). Substituting a lower quality make for the top-of-the-line car which was originally desired is equivalent to relaxing the cleanup criteria at a site because the money and/or technology doesn't exist to achieve the original target. Relying on natural attenuation for site cleanup is equivalent to deciding that, in fact: (1) your existing car isn't all that bad for your needs, and/or; (2) the newer cars offer little improvement.

Unlike the car case, the decision analysis for site cleanup must rely on analytic and numerical (computer) models. Despite the uncertainties inherent in applying models for contaminant reaction and transport within engineered and natural systems, there exists no better way to: (1) project into the subsurface or into the future; (2) evaluate the performance of various potential remedial alternatives; or (3) evaluate the effects of uncertainty in site parameters (e.g., contaminant amounts and types, soil properties and hydrologic properties, and processes — e.g., pathways for contaminant movement). At every iteration of a total system performance analysis, both a conceptual and quantitative understanding of natural attenuation is required, primarily because natural attenuation provides a "no action" baseline for comparing remediation strategies. During remedial action the natural attenuation factor must be understood in order to represent accurately the natural containment system component of many remediation strategies.

Risk and cost should be reduced by probabilistic and total systems analysis approaches because: they lead to selection of remediation strategies that reduce risk; they eliminate those that do not; and they minimize the amount of unnecessary data collection.

RBCAs

Assuming the ultimate object of the environmental cleanup is to minimize health risk in a cost-efficient and timely fashion, cleanup decisions and site characterization should reflect the interplay between cost and health-risk reduction. Cleanup targets must be a part of the balancing act because both cost and risk reduction depend directly on how "clean" is defined for a given site, or collection of sites. A number of attempts have been made to define protocols for the cleanup decision-making process. The most notable of these is the RBCA approach. Recognizing that all sites are different, RBCAs rely on a tiered approach of graduated response in site characterization and remediation efforts to soil and groundwater contamination. Initial site screening work places individual sites in specific tiers, depending on the potential risk and the specific characteristics of the site. Sites in the lowest tier pose little risk. The sites which pose the greatest health risk reside in the highest tier, where remediation efforts are most directly concentrated. Site characterization efforts are specifically directed at collecting only the information which is needed for risk-based decision making. Because these become extensive only at the limited number of sites which pose a large risk, less expense arises from extensive site characterizations of low-risk sites.

RBCAs are increasingly being used by the states in LUFT cleanups. As of August 8, 1996, states having RBCA programs in place included Georgia, Michigan, Oregon, South Carolina, South Dakota, and Texas. Of these, Michigan, Oregon, and Texas were considering expanding the approach beyond LUFTs. Alabama, Arkansas, Idaho, Illinois, Indiana, Iowa, Kansas, Kentucky, Nebraska, New Mexico, New York, Ohio, Oklahoma, Mississippi, Pennsylvania, Tennessee, and Utah were in the process of incorporating an RBCA approach. Of these, all but Arkansas, Iowa, Kansas, Kentucky, Nebraska, New Mexico and Tennessee were considering applying the approach to non-LUFT sites. All of the remaining states were at some point in incorporating the ASTM RBCA approach. (However, Maine, Maryland, Massachusetts, Nevada, North Carolina, North Dakota, and Vermont have not scheduled RBCA training.) (This listing came from a bioremediation newsgroup posting from Chuck Newell, Nov. 4, 1996.)

The states are buying in to RBCAs for the very simple reason that the alternative, engineered cleanup to background levels rapidly depletes state funds while often providing no clear health benefit. An RBCA approach seeks to calibrate cleanup efforts to reflect human and environmental health risk. The more pressing the health threat, the more urgent is the need for cleanup. There is a particular focus on fixing cleanup levels to fit potential future uses of a site. This is certainly a better approach than the present one which generally assumes that the future use of all aquifers will be as sources of drinking water. There is a strong push for an RBCA approach to be applied to CERCLA actions in the future. It is therefore important to outline the salient features of RBCAs. Figure 9.1 illustrates an RBCA decision tree, and the three-tiered approach to cleanup.

In tier 1 the type of contaminant present and the threat a site poses to human health is assessed. Generally this will involve source characterization (historical records of site activities and releases, maximum contamination levels of most toxic and mobile species, which media are contaminated), potential for exposure and degradation of beneficial uses (identification of receptors, determination of any impact on future use), extent of migration (sampling of nearby wells or sewer lines to roughly limit the contaminant travel distance). Sites are then pigeon-holed as being immediate, short-term, or long-term (>2 years) threats. A decision is made to move upwards into tier 2 if the calculated risk exceeds some reference value of risk-based screening level (RBSL). ASTM (American Society for Testing and Materials, 1995) shows how RBSLs can be calculated, i.e., how to generate a look-up table (note that the RBSLs may vary from state to state). RBSLs are based on a series of exposure pathways analyses that use a variety of physical parameters (Henry's law constants, diffusion coefficients, solubilities, etc.) and exposure parameters (soil ingestion rate, inhalation rate, cancer slope factors, etc.). If the calculated risk is less than the reference value, there is removal of the source, but otherwise little further is done with the site. There is no extensive site characterization effort.

In tier 2 the site is characterized further to determine if site-specific contaminant target levels (SSTLs) are exceeded. Again, the targets are based on rather straight forward hydrogeochemical assessments of contaminant attenuation. If the contaminant

The Present

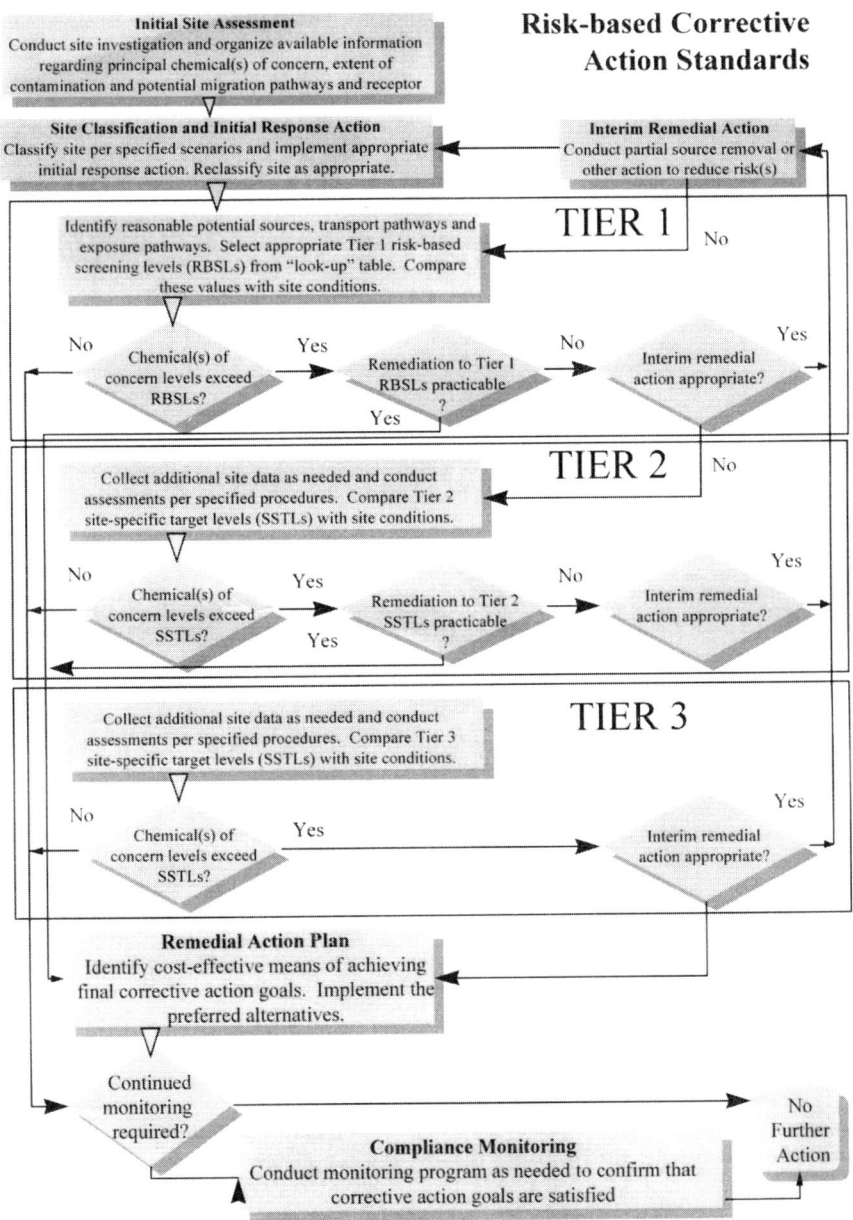

FIGURE 9.1 Risk-based corrective action decision tree. (From Begley, R., *Environ. Sci. Technol.*, 30, 438a–441a, 1996. Copyright 1996 American Chemical Society. With permission.)

levels do not exceed the targets, long-term monitoring is called for. If contaminant levels do exceed target levels, further site characterization is done in tier 3, this time with the possible object of actively implementing remediation strategies. Tier 3 involves an extensive assessment of potential fate, transport, and risk. It should be noted that the protocol of Wiedemeier et al. (1995b) for remediation of fuel hydrocarbons lends itself to RBCA's three-tier approach. Steps 1 and 2 of (Wiedemeier et al., 1995b) are analogous to the tier 1 stage of the ASTM RBCA. Steps 3 to 6 are analogous to tier 2, and Steps 7 and 8 would fall into tier 3.

Obviously, much depends on what the various states use in their look-up tables. If the RBSLs are set too high, all sites will remain in the lowest tier, tier 1. If the RBSLs are too low, all sites will move upward into the tiers requiring more intervention, and any savings from RBCA will be minimized. At this point, it is too early to tell on a state-by-state basis the overall direction of the RBCA process. RBCAs are useful because they force regulators and states to make explicit choices about the level of cleanup they want to pay for. There are problems though. RBCAs do not explicitly take account of environmental risks, i.e., non-public health risks. Moreover, politically, it may be difficult to incorporate an RBCA approach into CERCLA because RBCAs often result in the relaxation of cleanup goals. Until RBCA has been explicitly added to CERCLA, the cleanup will not change in scope or direction, as regulatory embrace of RBCA does not change the requirements of CERCLA.

Natural attenuation is in the ASTM RBCA guide for petroleum release sites (American Society for Testing and Materials, 1995). The ASTM RBCA guide notes how important the biodegradation sink is, followed by a listing of the range of measured biodegradation rates (ASTM is putting together a protocol for natural attenuation of petroleum release sites. The draft was put out for comment in March of 1996, and the final standard might be expected by the summer of 1997 (Ritz, 1996)). An appreciation of natural attenuation would tend to push sites to lower tiers. For example, the results of the Rice et al. (1995) report out of LLNL in essence said that most LUFT cleanups should remain in tier 1. If natural attenuation achieves greater acceptance, there will be less need in the RBCA approach for the politically unpalatable relaxation of cleanup targets.

RBCAs FOR METALS

Recent work (Jones, 1996a) has argued that an RBCA approach might be applied to the cleanup of metals-contaminated sites as well. Historically, metals in soils have been treated as if they were immediately available to groundwater and the biota. Consequently, remediation generally involves (Jones, 1996a): removal, cement-based stabilization, soil flushing, and ex-situ soil washing. The evidence cited in the preceding chapters suggests that many metals are quite inert in soils without the extra effort. Jones cites the following data needs for roughly assessing the risk posed by contaminant metals in soils:

1. Abundance of each metal species in each matrix present at the site (e.g., soil, water, sediment)
2. pH
3. E_H
4. Toxicity information and bioavailability data for the individual metal species

5. Other cations
6. Ion exchange capacity of the soil
7. Buffering capacity of the soil
8. Groundwater flow and direction

A conceptual site model describing the hazard source, exposure pathways, and receptors at a site, is useful in the RBCA approach, because it helps limit the amount of needed site characterization. Roughly constraining the chemical state of a contaminant metal in soils, also leads directly to an assessment of the risk a contaminated site poses. It should be noted that the state of Illinois is in the process of considering an RBCA approach for metals contaminated sites. The proposed rule can be downloaded from

(http:/www.state.il.us/pcb/proposal/742p.htm)

ENDNOTE

Because natural attenuation seems to be the appropriate site-treatment technique for a very large number of sites, it is prudent to ask "Is remediation by natural attenuation a bad choice?" as opposed to "Is natural attenuation an effective remedy?" Currently, natural attenuation is often relied on only if nothing else works. In the future we expect it will be considered at most sites before other methods are considered. Although this approach seems justified on the evidence presented above, there are certainly sites where natural attenuation will not work. If natural attenuation is mistakenly applied to a site where it is not appropriate, cleanup costs may escalate, and additional third-party liabilities may be imposed (e.g., Terauds, 1996). (Also, most state natural attenuation protocols require notification of nearby property owners; which will be unpalatable to some responsible parties). There should be a clear understanding of those cases where natural attenuation won't work. The AFCEE and RTDF protocols for natural attenuation of chlorinated solvents both have explicit decision points and grading criteria which allow the no-natural attenuation decision to be made. The same is true of the AFCEE fuel hydrocarbon protocol. So far, 45 Air Force sites contaminated with fuel hydrocarbons have been slated for remediation by natural attenuation.

Obvious site characteristics which should stop consideration of natural attenuation, or at least trigger further study, are close receptors and/or fast travel times. Under this umbrella fall sites which have high hydraulic conductivities and/or fast hydrologic pathways, and sites where drinking water wells are very close to the plume. Natural attenuation of chemicals which sorb poorly, are soluble, and are biodegraded slowly (if at all) should be of particular concern. Mercury and Cr(VI) under oxidizing conditions are two good examples of this. At sub-neutral pH (and non-reducing conditions) Cd, Pb, Co, and Zn fall into this category as well. Natural attenuation should be very critically examined if proposed for sites where the source term has not, or cannot be removed (DNAPLs). In many cases remediation by natural attenuation may require a somewhat longer period of time for site closure.

10 The Future

Environmental remediation is changing rapidly in this country because there is not enough money presently to support the environmental infrastructure that was erected by Congress, attorneys, environmental consultants, and regulators in the wake of CERCLA, and to pay for cleanup as well. In effect an "Environmental-Industrial Complex" is being paid large amounts of public and private funds to appear to remedy what may be a relatively minor threat to human and environmental health. There is very little push to pay more funds, which is just as well. More money will only sustain the edifice. So, does maintaining the status quo promise anything at all for the future? The optimist would point to the future development of innovative technologies. After all, Congress did the same thing when it passed CERCLA 16 years ago. It hasn't happened. The most innovative cleanup technology to gain regulatory acceptance in 26 years since CERCLA was passed was recognition of natural attenuation of petroleum hydrocarbons, as popularized in the Lawrence Livermore Report. And, despite recent studies which have revised downwards the estimated cost of Superfund cleanups (see Long, 1996) due to a slowdown in NPL listings, the total site cleanup is still expected to run in the hundreds of billions of dollars.

We believe the future of environmental remediation lies in a recognition that money matters and that results matter even more. But how to turn these fairly uncontroversial principles into a principled solution? Applying the concepts from the nation's first major environmental statute, the National Environmental Policy Act (NEPA) is a good start. NEPA applies to federal agencies whenever they approve a project that has the potential to adversely affect the environment. Where NEPA applies, a project proponent must prepare an environmental impact statement (EIS) that looks first at the project and its reasonably foreseeable impacts, then analyzes alternatives to the project. The alternatives analysis is at the heart of NEPA (as well as a number of other environmental statutes) and is intended to help focus the overall analysis of the original project's impacts. The current environmental remediation framework could never survive a searching NEPA analysis. We believe that in the future it must, and that under such an analysis, natural attenuation will become the first remedy considered. Money will be spent only where it addresses real threats and only where it will result in a corresponding public health benefit.

The first step in the NEPA process is the definition of the project. The alternatives analysis flows from the project definition. Save for the no-project alternative, all alternatives must achieve the goals of the project. Here, the project is the cleanup of a given site. Unfortunately, the degree of the cleanup is defined under Section 121 of CERCLA as:

1. Protective of human health and the environment
2. Relevant and appropriate under the circumstances
3. Will not exceed standards of more stringent state and federal laws, including MCLs

Only EPA decides which criteria applies in a given situation. And, under CERCLA, EPA always has discretion to change the applicable standard. In other words, the project is never defined. This is unfortunate because without a fixed project definition it is impossible to engage in a meaningful analysis of alternative methods to achieve the project and impossible to identify the most cost-effective alternative. At the same time, a moving project definition conceals the sad truth about the majority of CERCLA cleanups — they do not achieve many, much less all, of the optimistic goals Congress included on its CERCLA cleanup wish list.

The continued devotion of limited resources to meeting aspirations without limits results in the loss of credibility and failure to come to grips with real problems. Returning again to the car-buying analogy, if you were to enter the market for a new automobile and limit your search to automobiles that carried six people safely, reached a top cruising speed of 150 miles per hour in 15 seconds from a standing start, was safe, got 65 miles to the gallon, and looked good to boot — it wouldn't take too many sales lots before you would revisit your list of criteria and decide which of them were most important. Instead, in the cleanup field we keep buying the expensive models with the shiny mirrors and expecting that with enough time and money they will eventually become the machines we sought in the first place. Congress needs to define the project. If the project is to protect human health, cleanups of non-drinking water aquifers should be rare. If the project is to protect the "the environment" or a "resource" Congress needs to be clear which parts of each it can afford to clean up, because we simply can't print money fast enough to remove every single contaminant molecule from the environment. By way of illustration, a common problem in the Central Valley of California and other areas that rely on shallow aquifers for drinking water is chlorinated-solvent contamination from drycleaners (Wall Street Journal, California Edition, Oct. 3, 1996). No one disputes the fact that this contamination poses a threat to public health, but because the project is defined as something more than public health, the solution becomes a moving target and the cost estimates become unreal. The most effective way to remove the public health threat is to treat the contamination at the wellhead, typically with the use of granulated activated carbon filters ("GACs"). GACs are expensive but nowhere near expensive as cleanup of the aquifer, which is required under CERCLA and the NCP. In Lodi, California, a town of 50,000 in the Central Valley with six impacted wells, the cost to protect public health is 3 million dollars for six GACs (+ the expense of operating them). "Cleanup" of the aquifer is estimated to cost around 30 million dollars over 50 years. This figure does not include the cost of the six GACs which will be required in any event. Nor does it take into account the fact that no one from the regulators on down expects the cleanup to be successful because of the existence of DNAPL. In short, if the project is defined by reference to public health, the price tag is 3 million dollars (+ the expense of operating the GACs). If the definition includes "the resource" or "the environment" the price tag jumps an order of magnitude with little prospect of success, and with no corresponding public health benefit.

Even if Congress does define the project in terms of protecting public health, it should still require a thorough alternatives analysis as required under NEPA. This would serve a number of constructive purposes. First, because proposed alternatives must meet the same general goal of the project itself, the process would help focus definition of the project. Without a no-project alternative to set a baseline, there is no way to measure the efficiency or success of an engineered solution. Second and more importantly, a full alternatives analysis would include a "no-project alternative" and this would result in full consideration of natural attenuation.

EPA is already increasing its recognition of natural attenuation. At the time of writing (early 1997) EPA is in the process of putting together a natural remediation policy (Cooney, 1996). The main objectives of the policy are to: (1) Define what EPA considers natural attenuation, and (2) outline the supporting documentation EPA will expect for natural attenuation from site to site. The policy is not expected to include step-by-step protocols, though. The NCP, with its emphasis on quick, expensive fixes, guaranteed that EPA would seldom select natural attenuation as the method of remediation at an NPL site. See, for example, Borgert's (1995), note 2 wherein the author discusses his talks with various EPA regions about the use of natural attenuation. "Those EPA regions that will consider soil half-life in the risk-assessment process stated that their consideration would be on a site-specific basis, would require quality data to verify that degradation was occurring, and would limit the extent of extrapolation (i.e., that extrapolation to a 30-year exposure duration might reasonably require 15 years of data)." Recent history indicates that there has been a change. A review of the records of decisions (RODs) from the NPL from 1984 to 1991 indicates that natural attenuation was seldom selected as the final remedy. Beginning in 1992 a number of sites were delisted from the NPL each year, based upon the expectation that natural attenuation would remediate some or all of the contamination at a site. The following is a list of some NPL sites where natural attenuation formed part or all of the site remediation.

1991

1. Sheridan Site, Waller County, TX. Region 6.

1992

1. Western Sand and Gravel Site, RI, Region 1.
2. Kin-Buc Landfill, NJ, Region 2.
3. PSC Resources, MA, Region 1.
4. Twin Cities Air Force Reserve, MN. Region 5.
5. Town Garage/Radio Beacon (Holton Circle), NH, Region 1.
6. Mosley Road Sanitary Landfill, OK, Region 6.

1993

1. Juncos Landfill, Juncos County, Puerto Rico, Region 2.
2. Monroe Township Landfill Site, NJ. Region 2.
3. Ringwood Mines/Landfill Site, NJ, Region 2.

1994

1. Coakley Landfill, NH, Region 1.
2. Rockingham Landfill, Windham County, VT, Region 1.
3. Davie Landfill, Broward County, FL, Region 4.
4. B & B Chemical Company, Inc. FL, Region 4.
5. Dover Air Force Base, DE, Region 3.

1995

1. Wilson Concepts Site, Pompano Beach, FL, Region 4.
2. Diamond Shamrock Site, Polk County, GA, Region 4.
3. Jackson Township Landfill Site, Ocean County, NJ, Region 2.

1996

1. New Castle Spill Site, New Castle, DE, Region 3.
2. Alcoa (Vancouver Smelter) Site, Clark County, WA, Region 10.
3. McChord Air Force Base Washrack Treatment Area, Tacoma, WA, Region 10.
4. Oak Grove Sanitary Landfill, Anoka County, MN, Region 5.
5. Agrico Chemical Site, Pensacola, FL, Region 4.

As is often pointed out by regulator and consultant alike, every site is different. Nevertheless, the listed sites do have one element in common. By placing them on the NPL it was determined that they were among the worst sites in the country. None of these sites had "simple" problems capable of easy solution. They include, for example, mining waste, heavy metals, chlorinated solvents, volatiles, semi-volatiles, petroleum hydrocarbons, and coal combustion waste. We include extensive descriptions of each in Appendix III.

The increasing reliance on natural attenuation for these sites would be encouraging, except that it still takes immense amounts of time to ultimately decide that natural attenuation can remediate a given site. The RODs do not tabulate costs incurred during the multi-year period by attorneys, regulators, and environmental consultants. That is unfortunate because those costs would reflect the money lost because natural attenuation is not the first remedy considered but instead, the last resort. It is up to Congress to reverse the bias toward engineering projects. EPA can't reverse its emphasis and look first to natural attenuation and then to expensive engineering solutions, because of the NCP. Nor can EPA simply revise the NCP in a way which would be contrary to the statutory priorities set in CERCLA. The explicit recognition of natural attenuation in cleanups may provide the only principled avenue for Congress to correct the problem. Natural attenuation has to be forcibly given pre-eminence in existing environmental statutes. Natural attenuation isn't relied on because of the institutional prejudice for physical solutions to biogeochemical problems, and this can only be altered by changing the way environmental cleanups are looked at.

The project definition of a given cleanup should begin with public health and should be consistent with a realistic consideration of actual health receptors in soils and groundwater. Then a remediation method should be chosen that removes any pressing health risk immediately, and addresses any long-term threats to soil or groundwater over a time frame commensurate with the actual risk posed by the contamination. EPA already does the first. One of the remediation alternatives should always be the natural attenuation alternative, even if evidence gathered during site characterization indicates that this alternative will take longer or won't work at all. The fact that natural attenuation does not work in a given situation doesn't mean that some other method will. As we argue elsewhere, if the method doesn't work, it shouldn't be implemented. Currently, technologies that don't work are still required as if the payment of money at the surface automatically improved conditions below.

Where natural attenuation will work, it should be selected as the Final Corrective Action Method. Where natural attenuation will not work, and another method will, the other method should be selected as the Final Corrective Action Method. Where natural attenuation will not work and neither will any other method, the responsible regulatory agency should consider deed restrictions for the site in question and some useful form of punitive sanction such as payment into the Superfund or funding of public water treatment. Pumping and treating to clean contaminated aquifers that won't be clean in hundreds of years is a particularly useless punitive sanction.

CONCLUSION

There was an old woman who swallowed a fly … and, to catch the fly, moved on to progressively more drastic measures. Had she simply passed the fly (and avoided swallowing flies in the future) there would be no nursery rhyme and the authors of this book would have been deprived of perhaps the best metaphor for this nation's approach to the remediation of hazardous substances. CERCLA proceeds upon a very basic but very flawed premise — a hazardous substance in the soil or groundwater posesses immortal toxicity. If this premise were true, it might make sense to maintain the current battery of environmental consultants, lawyers and regulators. Or … a spider to catch the fly, a bird to catch the spider, a cat to catch the bird, a dog to catch the cat, etc. As we have explained in the preceeding chapters, hazardous substances are just as subject to the laws of nature as any other substance, and time works very drastic changes on the chemical reactivity of the majority of them.

CERCLA is under attack because it costs a great deal and gives very little subsequent public health benefit. Cleanup up the majority of the CERCLA-controlled sites to MCLs in a few short years is a battle that cannot be won, because, even if the money was available, the science and technology are not. The technology to wave the contamination away does not exist — it has not for the past two decades and probably won't for the next three. It is bad that soil and water was polluted by the introduction of hazardous substances. Spending a trillion dollars will in itself make them only marginally less polluted, and nowhere near their precontaminated states. CERCLA's public health aspirations must be preserved, and the only way to do that is to address public health receptors first, followed by judicious use of current scientific knowledge to address all other risks in the most cost-effective manner possible. In many cases this will result in selection of natural attenuation as the final remedy.

Appendix 1 — State Treatment of Natural Attenuation

Ritz (1996) examined, through an informal survey of state regulators, the application of natural attenuation as a stand-alone remedy by the various state environmental quality agencies, focusing on unpermitted sites with VOC impact to groundwater. Ritz grouped state responses into four categories; those states having informal policies (11); those having a written policy (8); those states where natural attenuation is implicit in their policies (21); and those who don't have a policy but would consider natural attenuation as a remedy (37). This adds up to more than 50 states because some states have multiple agencies working on groundwater cleanup. Consequently, some states show up in multiple categories because cleanup responsibilities are treated differently for different contaminant types. For example, Montana's non-LUFT section would consider natural attenuation, though there are only informal guidelines for its utilization with LUFTs.

The various groups are listed below.

1. Written policies — Minnesota (LUFT), Michigan (Environmental Response Division), Kentucky (LUFT), South Carolina (LUFT), North Carolina (LUFT), Idaho (Remediation Bureau), New Jersey, and Wisconsin. The latter three are using draft material pending final regulatory approval (Ritz, 1996). The policies of Kentucky, Idaho, Minnesota, New Jersey, and South Carolina deal strictly with natural attenuation of petroleum hdyrocarbons. Those of Michigan, North Carolina, and Wisconsin apply to chlorinated solvents as well. Site assessment in a few states (Kentucky, Minnesota, and New Jersey) requires measurement of indicator parameters (dissolved oxygen, TOC). North Carolina and Wisconsin emphasize historical trends and temporal changes in plume morphology.

- Kentucky — Responsible parties must publish notice of corrective action in a local newspaper. Monitoring required until asymptotic contaminant levels can be demonstrated.
- Michigan — Addresses institutional control. Requires monitoring until statewide groundwater standards are achieved.
- Minnesota — A minimum of six to eight quarters of monitoring required.
- New Jersey — Requires responsible party to obtain information about future use. A minimum of six to eight quarters of monitoring required. Requires monitoring of wells in the source area, between the source area and the plume fringe, and no less than three year time-of-travel upgradient of the nearest receptor and five years time-of-travel downgradient of the plume margin.

- North Carolina — Responsible party must notify all potentially affected and adjacent landowners (and some local officials) individually. Requires monitoring until statewide groundwater standards are achieved (See below).
- Wisconsin — Requires responsible party to obtain information about future use. Addresses institutional control. Off-site affected property owners must consent to groundwater restrictions.

North Carolina's natural attenuation policy is among the most comprehensive of state policies and requires the following:

(http://www.gvi.net/soils/JanFeb96/northcaro.html)

1. Completion of a comprehensive site assessment which determines, among other things, the horizontal and vertical extent of the plume
2. Removal of all contaminant sources (e.g., free-phase product)
3. An exposure assessment outlining the physical and chemical characteristics of the contaminant, as well as its toxicity and persistence, and potential exposure pathways
4. A cost comparison of the various potential remedial actions
5. Demonstration that the contaminant has the capacity to be degraded under site-specific conditions to North Carolina groundwater quality standards "within an acceptable period of time"
6. Demonstration that contaminant migration can be predicted
7. Demonstration that surface or groundwater standards will not be exceeded at receptors
8. Demonstration that migration onto adjacent properties will not occur, or that the adjacent property owner have consented in writing to the correction action plan, or that the latter properties possess alternative water sources
9. Implementation of a groundwater monitoring plan (see above)
10. A public comment period of 30 days once the correction action plan has been submitted
11. Certification of the correction action plan by a registered Professional Geologist or Professional Engineer

2. Informal policies/guidelines — Arkansas, Mississippi, Missouri, Ohio (Voluntary Action Program), Maryland (Hazardous Waste Program), Delaware (LUFT), Montana (LUFT), Arizona (LUFT), California (Dept. Toxic Substances Control), Wyoming (LUFT), North Dakota.

- Arizona — Only if monitoring demonstrates degradation, and if statewide water quality standards are met at the site boundary.
- Arkansas — Plume may migrate off site. Sites may be closed without monitoring if low levels of BTEX and no receptors.

- California — All parameters affecting natural attenuation must be analyzed.
- Delaware — Horizontal and vertical extent of plume must be defined first. Plume must stay on site.
- Montana — Horizontal and vertical extent of plume must be defined first. Plume may migrate off site.
- North Dakota — Requires monitoring for minimum of 2 years. Has approved implementation of natural attenuation at over 200 petroleum sites and 20 industrial solvent sites.

3. Implicit in policies/guidelines — Oregon (LUFT), Utah (LUFT), Colorado (LUFT), New Mexico (Remediation Program), South Dakota (LUFT), Oklahoma, Iowa (LUFT), Illinois, Michigan (LUFT), Ohio (LUFT), Florida (Dry cleaning Program/Engineering and Technical Support), Georgia (LUFT), South Carolina (non-LUFT), Pennsylvania (Environmental Cleanup Program), Delaware (Site Investigation/Restoration Branch), Massachusetts, Connecticut, Rhode Island, Vermont, New Hampshire, Maine.

In many of these states the cleanup standard is set depending on current and future use of the groundwater. Cleanup standards are stricter for drinking water aquifers than they are for non-potable groundwater. Use restrictions may be applied until monitoring indicates attenuation in some states. Illinois and Michigan reference the ASTM RBCA protocol in their guidelines.

4. No policies, but would consider — Alaska, Washington, Oregon (Cleanup Policy and Program Development Section), California (Water Resources Control Board), Nevada, Idaho (RCRA/CERCLA Sections), Montana (non-LUFT), Wyoming (non-LUFT), Colorado (non-LUFT), Utah (non-LUFT), Arizona (Soil and Grounwater Standard Development), New Mexico (RCRA/LUFT Sections), Texas, Kansas, Nebraska, South Dakota (RCRA), Minnesota (Superfund), Iowa (Uncontrolled Sites Section), Missouri (non-LUFT), Arkansas (Hazardous Waste Section/Water Division), Louisiana, Mississippi, Alabama (Hydrogeology Unit), Tennessee, Florida (Hazardous Waste Section), Georgia (Hazardous Waste Section), Kentucky (non-LUFT), Indiana, Ohio (Superfund), West Virginia, Virginia, Pennsylvania (CERCLA/Hazardous Waste/Solid Waste), Maryland (Oil Control Program/Superfund), Delaware (Groundwater Protection/RCRA Programs), Maine (Hazardous Material/Solid Waste), Rhode Island, New York.

State by state specifics follow (Ritz, 1996):
- Alaska — Property boundaries limit application of natural attenuation.
- California — Only petroleum hydrocarbons; no solvents.
- Colorado — May require deed restriction.
- Florida — Property boundaries limit application of natural attenuation.
- Idaho — No sites approved yet.
- Indiana — Might consider natural attenuation of hazardous industrial solvents as well as petroleum hydrocarbons. No sites approved yet.

- Kansas — Might consider natural attenuation of industrial solvents as well as petroleum hydrocarbons.
- Kentucky — May require deed restriction.
- Louisiana — Might consider natural attenuation of industrial solvents as well as petroleum hydrocarbons.
- Maryland — Only for urban or industrial areas where no potential receptors exist.
- Mississippi — Allows natural attenuation only if fastest, or only, way to remediate site. Might consider natural attenuation of industrial solvents as well as petroleum hydrocarbons.
- Missouri — May require deed restriction. No sites approved yet.
- Montana — May require deed restriction.
- Nebraska — Only for urban or industrial areas where no potential receptors exist. Might consider natural attenuation of industrial solvents as well as petroleum hydrocarbons. RBCA + natural attenuation gaining favor.
- Nevada — Reluctant to use natural attenuation because of heavy reliance on groundwater for drinking. Might consider natural attenuation of industrial solvents as well as petroleum hydrocarbons.
- New York — Might consider natural attenuation of industrial solvents as well as petroleum hydrocarbons.
- Ohio — Sites threaten receptors, so natural attenuation unlikely to be implemented.
- Rhode Island — Natural attenuation can only be used in conjunction with active remediation (at the edges of a plume), or if the latter can no longer recover contaminants. Might consider natural attenuation of industrial solvents as well as petroleum hydrocarbons. 5% of sites have used natural attenuation.
- Tennessee — Allows natural attenuation only if fastest, or only, way to remediate site. No sites approved yet.
- Utah — May require deed restriction.
- Virginia — RBCA + natural attenuation gaining favor.
- Washington — Natural attenuation can only be used in conjunction with active remediation (at the edges of a plume), or if the latter can no longer recover contaminants. Might consider natural attenuation of industrial solvents, as well as petroleum hydrocarbons.
- West Virginia — Reluctant to use natural attenuation because of heavy reliance on groundwater for drinking. No sites approved yet.
- Wyoming — Has approved ten sites for natural attenuation.

To the categories noted above must be added those states which anticipate policy changes affecting natural attenuation (as of October, 1996), which is every state with the exceptions of Nevada, the Dakotas, Oklahoma, Iowa, Wisconsin, South Carolina, Connecticut, Maryland, Alaska, Hawaii, and Massachusetts. Policy changes are expected to be imminent in Oregon, New Mexico, Texas, Ohio, and Florida. In the others the existing policies are under review.

Appendix 1 — State Treatment of Natural Attenuation

Ritz (1996) noted that there is an increasing recognition of natural attenuation as a remedial option by all states. The movement of the states to formalized policies is, at present, largely focused on petroleum hydrocarbons. At the same time natural attenuation has been folded into RBCA policies. Emerging policies tend to emphasize primary lines of evidence to demonstrate natural attenuation. There is also an increasing acceptance of natural attenuation for chlorinated solvents. The states which permit natural attenuation require: source removal and control, a site assessment that defines the horizontal and vertical extent of contamination; and follow-up monitoring to ensure no impact to nearest receptors. Regulations vary in regard to duration of monitoring, the use of institutional controls, and what contaminants can be considered (Ritz, 1996).

Appendix 2 — Glossary

Aerobic respiration — Microbiological processes which use oxygen as an electron acceptor.

AMD — acid mine drainage.

Anaerobic respiration — Processes whereby microorganisms use a chemical other than oxygen as an electron acceptor.

Aquifer — A rock body which stores water.

Aquitard — An underground rock formation having low permeability and which retards groundwater movement.

Aromatic hydrocarbons — a compound built of benzene ring(s).

Bacterium — A single-celled organism of microscopic size.

Biodegradation — The breakdown of organic molecules into, ultimately, CO_2 and H_2O.

Biotransformation — Microbially engineered transformation of a chemical compound to another.

BTEX — Benzene, toluene, ethylbenzene, and xylene.

CERCLA — The Comprehensive Environmental Response, Compensation and Liability Act, 42 U.S.C. Section 9601 to 9675. Passed by Congress in 1980 and amended numerous times since then, most drastically in 1986 with the passage of SARA, the Superfund Amendments and Reauthorization Act. CERCLA contains numerous provisions including a liability scheme and the Superfund.

CFR — The Code of Federal Regulations is the body of regulations promulgated by the various federal agencies implementing statutory mandates from the United States Congress. CFR provisions have the force of law.

Clean Water Act — The Federal Water Pollution Control Act, 33 U.S.C. sections 1251 to 1387. Passed by Congress originally in 1948 and amended numerous times since then, most drastically in 1972.

Cometabolism — Microbial transformation of one compound that requires the metabolization of another compound, the latter termed the primary substrate.

Diffusion — The movement of a chemical from areas of higher concentration to lower concentration.

Dispersion — The "smearing" of a chemical due to a mixture of chemical diffusion and mechanical mixing.

DNAPL — Dense, non-aqueous phase liquid.

DOE — The United States Department of Energy.

DOD — The United States Department of Defense.

EPA — The United States Environmental Protection Agency.

Federal Register — Regulations promulgated in draft and final form by federal agencies pursuant to statutory mandates. Only final regulations in the Federal Register have the force. The Code of Federal Regulations is updated periodically to include final rules from the Federal Register.

Half-life — The amount of time required for one half of the total amount of a given chemical to be transformed to another through radioactive decay, biodegradation, or hydrolysis.

Halogenated hydrocarbons — Organic molecules possessing a halogen (Cl, Br, F).

Hydraulic conductivity — A measure of the permeability of an aquifer to fluid flow.

Hydraulic gradient — The drop in aquifer potentiometric surface over a specified distance.

Intrinsic bioremediation — The reduction in contaminant mass through degradation by indigenous microorganisms in the absence of engineered help.

Intrinsic Remediation — See natural attenuation.

Isotope — An element having a specific number of protons and neutrons in its nucleus.

Joint and Several Liability — When a creditor or enforcing party may bring a legal action against one or more of a class of responsible parties for the full cost of an obligation. CERCLA imposes joint and several liability.

LNAPL — Light, non-aqueous phase liquid.

LUFTs — An acronym created by EPA to describe Leaking Underground Fuel Tanks. Replaced the earlier acronym "LUST" describing Leaking Underground Storage Tanks.

MCL — Maximum Contaminant Level. An acronym used most often in connection with the Safe Drinking Water Act.

NAPL — Non-aqueous phase liquid.

Natural attenuation — The umbrella of natural processes which break down contaminants and decrease their health risk.

NCP — The National Contingency Plan for the Removal of Oil and Hazardous Substances, promulgated by EPA pursuant to the direction provided by Congress in Section 121 of CERCLA. 42 U.S.C. Section 9621.

NEPA- The National Environmental Policy Act, 42 U.S.C. sections 4321 to 4370d. Passed by Congress in 1970. NEPA is a statute requiring federal agencies to perform a searching analysis of the environmental effects of acts performed by or authorized by federal agencies.

NORM — Naturally occurring radioactive matter (e.g., uranium, thorium, radon).

NPL — The National Priorities List. A list of the most contaminated sites in the United States listed by EPA pursuant to Congress's direction in Section 105 of CERCLA, 42 U.S.C. Section 9605.

NWPA — The Nuclear Waste Policy Act of 1982. The NWPA identifies the parties responsible for the disposal of high-level radioactive waste in the United States.

Oxidation — The donation of electrons by a chemical compound to an electron acceptor.

PAH — Polycyclic aromatic hydrocarbons.

PCB — Polychlorinated biphenyl.

Persistence — The tendency for a given contaminant to resist chemical transformation.

Appendix 2 — Glossary   **181**

Plaintiff — The person who brings a legal action.

Plume — The three-dimensional zone of contamination.

Point of compliance — A location between the source area and a potential exposure point where contaminant levels must be below the remedial target levels.

Radioactive Decay — The process whereby atomic nucleii break down/transform to form new isotopes isotopes.

Receptor — Person, structure, well, or water supply that might be affected by a release of contaminants.

Reduction — The accepting of electrons by a chemical compound.

Reductive dehalogenation — Microbially catalyzed replacement of a halogen by hydrogen.

RCRA — The Resource, Conservation and Recovery Act, actually part of the Solid Waste Disposal Act, 42 U.S.C. 6901 to 6992k. Enacted in 1976, RCRA set up a "cradle-to-grave" management system for hazardous wastes. As such it contains provisions dealing with management of hazardous wastes at landfills and other disposal facilities.

Retroactive Liability — Retroactive liability is created by a statute which imposes liability for acts that did not give rise to liability before enaction of the statute in question. CERCLA and RCRA impose retroactive liability.

SARA — The Superfund Amendments and Reauthorization Act. Passed in 1984, SARA amended both CERCLA and RCRA.

Saturated zone — The portion of a rock or soil body in which all of the pores are filled with water.

SDWA — The Safe Drinking Water Act, 42 U.S.C. 300f and following. Originally passed by Congress in 1944 and amended a number of times since then, most drastically in 1974.

Sorb — The taking up and holding of chemicals on solid surfaces.

Sorption — The sticking of an otherwise dissolved compound onto or into a mineral surface

Strict liability — Strict liability is imposed by a statute which imposes criminal and/or civil liability for an act without requiring a showing of specific intent.

Substrate — Compound used by microorganisms for metabolic functions.

Superfund — The Superfund was created by CERCLA and essentially set up a program where tax revenues raised under CERCLA would be devoted to the cleanup of some, but not all, of the contaminated sites in the United States.

TCLP — Toxicity Characteristic Leach Procedure.

Vadose zone — the portion of a rock body between the water table and the ground surface where the pores are not filled completely with water.

Volatilization — the partitioning of compounds dissolved or mixed in a liquid phase into the gas phase.

U.S.C. — The United States Code consists of laws passed by the United States Congress.

Appendix 3 — Excerpts From NPL Site RODs*

1991

1. Sheridan Site, Waller County, TX. Region 6.
EPA ID# TXD062132147
Date Listed on the NPL: 1986
Date of Final Remedy Selection: 1988

The Sheridan Disposal Services site is located about 9 miles northwest of Hempstead, and is bordered by the Brazos River and Clark Road. The site covers approximately 110 acres of this 695-acre tract of land and operated as a commercial and industrial waste disposal facility from 1958 through 1984. A 15-acre sludge lagoon, a 40-acre evaporation landfarm, nine storage tanks, and incineration plots were used for waste disposal. A pond levee around the lagoon was constructed, encompassing 17 acres. The State banned waste disposal in the lagoon in 1976, and revoked Sheridan Disposal Services' waste disposal permit in 1984 because the firm lacked technical and financial resources to adequately close the site. Elevated levels of heavy metals were found in river sediments downstream of the site. The Town of Brown College, with approximately 60 people, is about 1½ miles north of the site. The nearest residence and drinking water wells are less than 1 mile from the site. Land immediately surrounding the site is agricultural, including pasture and range lands.

The groundwater is contaminated with VOCs, including benzene, ethyl benzene, trichloroethylene (TCE), and toluene. The soil and sludge are contaminated with VOCs, including benzene and toluene, as well as PCBs. The upper aquifer, which is connected to the Brazos River, is contaminated and believed to be connected to the lower Evangeline Aquifer. The Brazos River, the shallow alluvial aquifer, and Evangeline Aquifer are used for drinking water supplies. Direct contact with contaminated soil is unlikely, since access to the site is limited. In 1978, water overflow from the site killed fish in Clark Lake, but off-site sampling of the Brazos River and Clark Lake from 1984 to 1986 detected no contamination. Marshlands lie 3,000 feet to the east of the site.

---

* These are excerpts from NPL site RODs where natural attenuation was selected as part of the final remedy. Sites were identified on the basis of a legal word search of RODs, and a posting on the bioremediation mail group by Robert Lunardini. The list below is probably incomplete as the actual number is thought to be between 50 and 100, though it depends on how the question is posed to EPA. For example, all landfill sites involve natural attenuation in that the contaminants remain in the landfill. The sites below probably don't include all of the sites where natural attenuation forms part of the final remedy (R. T. Hackler, personal communication, 1/24/97). A more extensive, updated list can be found at: http://www.sandia.gov/eesector/gs/gc/snaprod/htm

This site is being addressed in three stages: initial actions and two long-term remedial phases focusing on soil and sludge cleanup and groundwater treatment. In 1986, a fence was installed around the site to control unauthorized site access. Periodic maintenance of the levee system also has occurred to prevent flooding of former disposal areas and possible contamination of the Brazos River. In 1987, the potentially responsible parties, with EPA oversight, transferred rain-water to the evaporation pond for on-site treatment.

To control the source of contamination, the potentially responsible parties will, under supervision by the EPA, use bioremediation to reduce PCB levels in soil and sludges. Treated sludges will then be stabilized, returned to the pond, and capped. Treated sludges that still have elevated PCB levels will be disposed of in a federally approved landfill in the pond area. The cleanup was scheduled to begin in late 1995.

The EPA selected natural attenuation as the remedy for groundwater contamination. This remedy relies on natural processes such as sorption and biodegradation to alleviate contamination. Because groundwater moves so slowly, it is expected to take a minimum of 30 years for the contamination to be eliminated. The remedy provides for: monitoring of surface water to ensure that protective levels are maintained in the Brazos River, which would be the first point of potential exposure to contaminated groundwater; monitoring of groundwater to track movement of the contaminant plume; and prevention of future use of groundwater as a source of drinking water for nearby residents through deed restrictions and other precautions. The remedy also established contaminant concentration limits specifically for this site, including enforceable water quality measurements that are designed to ensure that no contamination is found in the Brazos River. Design of the remedy is currently underway; cleanup activities are scheduled to begin once the treatment for soil and sludges is complete.

In 1987, 58 potentially responsible parties entered into an Administrative Order with the EPA to conduct an investigation of the feasibility of various methods of cleanup. The Order was amended to include eight additional potentially responsible parties. A group of potentially responsible parties has formed the Sheridan Site Committee. Based on an Administrative Order on Consent, the committee agreed, in 1991, to conduct a pilot of the bioremediation technology. The pilot was successful. The initial actions to secure the site and to treat or contain liquid wastes and contaminated rainwaters have reduced exposure risks at the Sheridan Disposal Services site. The site is safe while final cleanup remedies are being designed.

1992

Western Sand and Gravel Site, RI. Region 1.
EPA ID# RID009764929
Date Listed on the NPL: September, 1983
Date of Final Remedy Selection: December, 1992

Western Sand & Gravel, a 20-acre site located in a rural residential area of Burrillville, was a sand and gravel quarry operation from 1953 until 1975. The quarrying operation continues today. From 1975 to April 1979, approximately 12 acres of the

Appendix 3 — Excerpts From NPL Site RODs

20-acre site were used for the disposal of liquid wastes, including chemicals and septic waste. Over time, the wastes penetrated into the permeable soil and contaminated the groundwater. Contents of tank trucks were emptied directly into 12 open lagoons and pits, none of which were lined with protective materials. The pits were concentrated on a hill that slopes to Tarkiln Brook, which is used for recreational purposes and drains into the Slaterville Reservoir. The State closed the disposal operation because nearby residents complained of odors. Approximately 600 people within a 1-mile radius of the site depend on groundwater. Eight homes were found to have contaminated wells.

The site is being addressed through Federal and potentially responsible parties' actions. On-site groundwater is contaminated with VOCs, including toluene, TCE, trichloroethane, benzene, chlorobenzene, and dichloroethane. The water of Tarkiln Brook contains similar contaminants. The soil also is contaminated with VOCs. Prior to the capping of the soil and sludge and the installation of carbon filters, potential exposure to VOCs may have occurred by ingestion or direct contact with contaminated soil or groundwater.

In early 1980, the State began to pump one lagoon dry to halt leachate movement. Approximately 60,000 gallons of liquid chemical and septic waste were removed for off-site disposal. A groundwater recirculation system was installed. The EPA built a permanent alternate water supply to service approximately 56 parcels of land. The potentially responsible parties installed carbon canister filters as a temporary protective measure in all the homes in the affected area until the permanent water supply was functional. Construction of the permanent water line was completed in 1992.

In 1988, the parties potentially responsible for contamination installed a cap over the areas of contaminated soil and sludge and graded the site to promote runoff and drainage. The site was also fenced and the potentially responsible parties agreed to maintain the fence, cap, and site. All construction is complete. The potentially responsible parties conducted an investigation to determine the extent of contamination and to evaluate alternatives for cleanup of the off-site groundwater. The investigation was completed in early 1991. Based on the investigation, the EPA selected a remedy of cleanup through natural attenuation. The site will be monitored through 1995. At that time, a system to pump-and-treat the groundwater will be installed if monitoring shows that natural cleanup is not occurring as predicted. If natural cleanup is working as expected, the potentially responsible parties will monitor groundwater and conduct evaluations every three years, with EPA oversight.

Approximately 45 potentially responsible parties entered into a Consent Decree with the EPA and agreed to pay for past costs, to build a cap, to conduct an investigation to determine the nature and extent of contamination, and to identify alternatives for cleanup of contaminated groundwater. The parties also agreed to pay the EPA for the cost of construction of the alternate water supply system.

Construction of all cleanup activities are complete, including fencing, capping, and grading the contaminated areas of the Western Sand & Gravel site, installing carbon canister filters, installing an alternative water supply system, and installation and monitoring of a groundwater monitoring network. Stabilizing the site and providing an alternate water supply system are keeping the site safe while natural processes clean the groundwater.

Kin-Buc Landfill, NJ, Region 2.
EPA ID# NJD049860836
Date Listed on NPL: September, 1983
Date of Remedy Selection: 1992

The 220-acre Kin-Buc Landfill site is an inactive landfill that operated from the late 1940s to 1976. From 1971 to 1976, the site was a State-approved landfill for both solid and liquid industrial and municipal wastes. The site accepted hazardous waste during this period, until the State revoked its permit in 1976 because of violations of several environmental statutes. An estimated 70 million gallons of liquid wastes, including 3 million gallons of oily waste, and over 1 million tons of solid waste, were disposed of between 1973 and 1976. The Kin-Buc site includes two major mounds, Kin-Buc I and Kin-Buc II, and one minor mound, Mound B. Site activities included burying and compacting contained wastes in Kin-Buc II and discharging hazardous liquid wastes into bulldozed pits at the top of Kin-Buc I. Three pits of black, oily leachate, designated Pits A, B, and C are located at an edge of Kin-Buc I. Adjacent to the pits is an impoundment referred to as Pool C. Oil, heavily laden with PCBs, accumulates in Pool C and then discharges into Edmonds Creek, a tributary of the Raritan River. The pond also holds leachate that contains chlorinated VOCs, which are believed to come from the landfill. The Edison Township Municipal Landfill lies 600 feet to the south of the site. A refuse-filled low-lying area is located between Kin-Buc I and the Edison Landfill. Three thousand people live within 3 miles of the site. The site is located in a wetlands area adjacent to the Raritan River.

PCB-contaminated leachate from the landfill has been seeping into an area known as Pool C. Elevated levels of PCBs were found in sediment samples in Edmonds Creek, Rum Creek, and the Raritan River. Contaminants seeping into the wetlands may harm wildlife. Elevated levels of PCBs in edible fish and shellfish have been detected. The food chain may be contaminated with PCBs, cadmium, and other heavy metals. Eating PCB-contaminated food may cause a wide range of ill effects in people. PCBs and a large number of other pollutants were detected in surface water. Concentrations of PCBs were found in shallow wells in the refuse layer of the site. The sand and gravel aquifer beneath the site is contaminated with leachate from VOCs and heavy metals. The bedrock aquifer may be contaminated. While in operation, frequent major fires and a number of serious occupational injuries have occurred at the landfill.

This site is being addressed in three stages: immediate actions and two long-term remedial phases focusing on cleanup of the two major mounds and Pool C, and cleanup of the adjacent waterways and wetlands. In 1980, the EPA began cleanup activities consisting of collection, treatment, and disposal of oily and aqueous phase leachate from Pool C. In 1982, as part of the settlement negotiations, the owners assumed responsibility for cleanup activities. In 1984, 4,000 drums containing oily and aqueous phases of leachate and contaminated solids were shipped off site for incineration. From 1984 to 1991, 3,123,755 gallons of aqueous phase leachate were shipped off site for treatment and disposal. As of 1991, 30,400 gallons of oily phase leachate had been shipped off site for incineration.

Two Major Mounds and Pool C: The cleanup technologies selected in the 1988 remedy to address these areas include: installation of a slurry wall on all sides of the site, collection and off-site incineration of oily phase leachate, collection and on-site treatment of aqueous phase leachate and contaminated groundwater with direct surface water discharge, maintenance and possible upgrading of the existing cap on Kin-Buc I, installation of a cap on Kin-Buc II and other portions of the site, including the Pool C area, long-term periodic monitoring, and operation and maintenance. The parties potentially responsible for site contamination, under EPA oversight, completed the technical specifications and design for the selected cleanup technologies in 1993. Construction began in mid-1993 and is scheduled to be completed in 1996.

The potentially responsible parties have completed the investigation of contamination in the wetlands, surface waters and groundwaters at and adjacent to the site. The investigation identified VOC's in the groundwater, as well as elevated levels of PCBs and metals in sediments and local wildlife, and concluded that the contaminated wetlands sediments are a source of the contaminants observed in local aquatic wildlife. In 1992, the EPA selected the excavation of wetlands sediments contaminated by PCBs and the disposal of these sediments within the landfill, as well as the restoration of excavated areas to address this portion of the site. The remedy selected for the major landfill mounds will be protective of the groundwater by preventing further release of leachate.

The potentially responsible parties finished the site design and began cleanup activities in the fall of 1994. The proposed cleanup activities are expected to be completed in 1996. The selected remedial action includes using natural attenuation to achieve groundwater cleanup levels.

PSC Resources, MA, Region 1.
EPA ID# MAD980731483
Date Listed on NPL: September, 1983
Date of Remedy Selection: 1992

The 21.5 acre PSC Resources site operated in the 1970s as a waste-oil refinery and solvent-recovery plant. The facility reclaimed drained oils and solvents from Massachusetts collection points, treated them with heat, and sold them as lube-oil base stock, road spray, and heavy-fuel mixes. Millions of gallons of waste were left behind in tanks and lagoons when the current owner abandoned the plant in 1978. After a spill in 1982, the EPA discovered several leaking tanks and containment dikes, as well as saturated soils. Surface waters, wetlands, and groundwater are directly threatened by the waste. Approximately 4,500 people live within 3 miles of the site. The Quaboag River is located 200 feet southwest of the site and is used for swimming and fishing. The property is located near a residential and commercial district and is adjacent to the Town athletic field.

Shallow groundwater contamination consists mostly of VOCs including benzene and methylene chloride. PCBs, including Aroclor-1248 and Aroclor-1260, and lead have been found in soil samples. The surface water and oil in the dikes contain the heavy metals arsenic and lead, as well as benzene and PCBs. Oil in a rainwater catch basin contains PCBs and tetrachloroethylene. Contaminants have been detected in the soils and shallow groundwater in the nearby wetlands. The site is located in a 100-year flood plain, providing conditions for flooding to wash contaminants from the site into the Quaboag River. People may be exposed to contaminants by inhaling polluted air emanating from the site, coming in direct contact with or accidentally ingesting contaminated water or soil, or by eating contaminated fish.

The site is being addressed in two stages: initial actions and a long-term remedial phase focusing on cleanup of the entire site. The tanks were emptied of over 1 million gallons of hazardous wastes between 1979 and 1984. In 1986, the State Department of Environmental Protection (DEP) cleaned and removed the tanks. The DEP also fenced the site in 1986. The EPA completed the repair and reinforcement of the fence in the fall of 1991. The repair was necessary to limit unauthorized access and to extend the fence to include the debris pile and spill area on the western and southern sides of the site. Warning signs also were posted along the fence and on facility buildings.

The DEP has studied the nature and extent of the contamination at the site. The investigation defined the contaminants and recommended alternatives for the final cleanup. The study was completed in May 1991, and a proposed plan for cleanup was distributed for public comment. The remedy, selected in 1992, calls for the use of in-place stabilization of the on-site contaminated soils and sediments, followed by capping. The engineering design of the remedy began in late 1994 and was scheduled for completion in mid-1996.

In 1982, acting under authority of the Clean Water Act, the EPA asked the owner to contain the oil discharge, determine the contents of 22 tanks, and investigate the possibility of groundwater contamination. The owner complied with all requests. In 1994, the EPA, the Department of Justice, and the State Attorney General's Office announced a settlement with approximately 165 potentially responsible parties who have agreed to pay $6 million to cover past costs and the cost of cleaning up the site.

The removal of hazardous wastes and installation of a fence have reduced the potential of exposure to hazardous substances at the PSC Resources area while further cleanup activities are being designed. The selected remedial action includes using natural attenuation to achieve groundwater cleanup levels.

Twin Cities Air Force Reserve, MN, Region 5.
EPA ID# MN8570024275
Date Listed on the NPL: July, 1987
Date of Final Remedy Selection: 1992

Since 1944, the 280-acre Twin Cities Air Force Reserve Base (Small Arms Range Landfill) was used for operations that resulted in the storage and disposal of hazardous substances. The Small Arms Range Landfill was the main base landfill from

1963 to 1972. The site is situated along the Minnesota River and covers approximately 3 acres. In addition to general base refuse, quantities of paint sludge, paint filters, and leaded-fuel sludge also were disposed of at the landfill.

The site is located within the 100-year flood plain of the Minnesota River and is periodically flooded, resulting in the release of chromium, lead, and zinc to the river. Approximately 64,700 people living in the Minneapolis-St. Paul metropolitan area depend on public and private wells for drinking water within a 3-mile area of the landfill. Monitoring wells showed contamination with low levels of mercury, chromium, lead, and zinc in the groundwater. Soil and sludge were contaminated with paint by-products and petrochemicals.

Immediate Actions: In the spring of 1987, the EPA secured the site, posted warning signs, transferred liquids to on-site storage tanks, shipped 69 drums of organic sludges for incineration, and transported 35 cubic yards of contaminated soil for off-site disposal. *Entire Site:* The Air Force completed an investigation of the site in 1992, determining the extent of contamination. The selected remedies included allowing contaminants to dissipate through natural attenuation, monitoring the groundwater, and imposing deed restrictions to limit site use. In addition, a fence has been constructed to secure the site. After 2 years of monitoring the groundwater, the site was re-evaluated to ensure that natural attenuation has been sufficiently reducing groundwater contamination. Cleanup goals have now been met and the site is currently being deleted from the NPL.

Site Facts: The Twin Cities Air Force Reserve Base is participating in the Installation Restoration Program, a specially funded program established by the Department of Defense (DOD) in 1978 to identify, investigate, and control the migration of hazardous contaminants at military and other DOD facilities. Environmental Progress All construction at the site is complete. The immediate actions, including the removal of liquid and solid wastes and contaminated soil, the construction of a fence, and the imposing of deed restrictions have eliminated the potential for exposure to hazardous substances at the Twin Cities Air Force Reserve Base (Small Arms Range Landfill) site. The groundwater has recovered through natural processes.

Town Garage/Radio Beacon (Holton Circle), N.H. Region 1.
EPA ID# NHD981063860
Date Listed on the NPL: March, 1989
Date of Final Remedy Selection: 1992

The Town Garage/Radio Beacon site includes a development of about 25 homes called Holton Circle, the Town Garage property on High Range Road, and an undeveloped woodland and wetland area. The site includes a series of residential wells and one commercial well, known as the Town Garage Well. The State conducted tests in 1984 and found several of the wells to be contaminated with VOCs. The EPA and the State have been investigating the area since 1985, and have not yet verified a source of the contamination. The Department of Defense owned the Town Garage property, located 1,000 feet west of the Holton Circle development, from the early 1940s to 1968 and operated a radio beacon there during World War II. The EPA also investigated a small auto repair shop located approximately 1,000 feet south of Holton Circle. The shop uses 1 to 2 gallons of degreasing solvents annually. The area around the site consists of mixed rural and residential properties and is being actively developed. Approximately 7,400 people obtain drinking water from private wells located within 3 miles of Holton Circle.

The site is being addressed through Federal and State actions. The groundwater in the wells is contaminated with volatile organic compounds (VOCs) including dichloroethylene and dichloroethane. The six residences with contaminated drinking water wells have been connected to a public water supply. In 1992, the EPA completed an investigation into the groundwater contamination at the site. The final cleanup remedy selected to address the contamination in the shallow and bedrock groundwater is to allow groundwater to clean itself through natural attenuation, along with institutional controls to prevent the use of groundwater for domestic purposes. This cleanup process could take from 7 to 25 years to complete. The State will conduct long-term groundwater monitoring.

After adding this site to the NPL, the EPA assessed conditions and determined that, besides connecting six residences with contaminated wells to the public water supply and preventing the use of contaminated groundwater for domestic purposes, no further actions are required to make the Town Garage/Radio Beacon site safe while natural attenuation of the groundwater occurs.

Mosley Road Sanitary Landfill, OK. Region 6.
EPA ID# OKD980620868
Date Listed on the NPL: February, 1990
Date of Final Remedy Selection: 1992

The Mosley Road Sanitary Landfill covers 72 acres and was used from 1975 to 1987 as a commercial, residential, and industrial landfill. In 1976, the landfill accepted approximately 2 million gallons of hazardous substances under a Temporary Emergency Waiver for Hazardous Waste Disposal issued by the Oklahoma State Department of Health. According to the permit application, pesticides, industrial solvents, sludges, waste chemicals, and emulsions were deposited into three unlined pits. Since then, the pits have been buried under as much as 80 feet of solid refuse and fill and a clay cap. In October 1984, Waste Management of Oklahoma, Inc. acquired the site, operating it until November 1987, when the landfill reached its authorized capacity. Concerns about groundwater contamination brought the site to the EPA's attention. Hazardous wastes were disposed of near the base of the landfill; a long-term risk could exist if wastes leak into the groundwater.

The landfill lies above the Garber-Wellington Formation, an aquifer that serves as a high-quality drinking water source for many Oklahoma City residents. The surrounding area is both residential and commercial. An estimated 57,000 people obtain drinking water from public and private wells within a 3-mile radius of the site. Six homes are located within 1/2 mile from the site and obtain drinking water from private wells. The soil and groundwater are contaminated with pesticides, industrial solvents, sludges, waste chemicals, emulsions, and other substances disposed of in the landfills. Potential contamination of the groundwater connected to the public drinking water system may pose a threat to public health.

The site is being addressed in two stages: initial actions and a long-term remedial phase focusing on cleanup of the entire site. In 1988, a clay cap was placed over the landfill. The site is also fenced to prevent unauthorized access. Under an agreement with the EPA, Waste Management of Oklahoma, Inc. and Mobile Waste Controls, Inc. conducted a study of the nature and extent of site contamination and evaluated potential remedies for site problems. The investigation was completed in 1991. In mid-1992, the EPA selected a cleanup remedy that includes: restoring groundwater as a potential source of drinking water through natural attenuation; continuing to monitor groundwater if samples indicate natural dissipation is ineffective; repairing and improving the existing cap and adding a vegetative soil layer; installing warning signs; restricting future use of the site; fencing the site; and implementing a landfill gas monitoring system to prevent explosions. The technical designs for these actions are underway and are expected to be completed in 1995.

Waste Management of Oklahoma, Inc. and Mobile Waste Controls, Inc. signed an Administrative Order with the EPA in 1989 to conduct an investigation into the nature and extent of site contaminants. In early 1994, the EPA issued a Unilateral Administrative Order to Waste Management of Oklahoma to clean up the site. In addition, the EPA is negotiating with over 40 potentially responsible parties to contribute to the cleanup efforts.

The site is now inaccessible due to the fence installed by Waste Management of Oklahoma, Inc. The installation of a clay cap over the landfill has reduced the potential for exposure to contaminants while final cleanup measures are being designed.

1993

1. Juncos Landfill, Juncos County, Puerto Rico. Region 2.
EPA ID# PRD980512362
Date listed on the NPL: September, 1983
Date of Final Remedy Selection: October, 1993

The 11-acre Juncos Landfill is a closed municipal landfill at which thermometers containing mercury were dumped. Small leachate seeps and soil erosion were evident during the site inspections conducted by the EPA. Of greatest concern were houses adjacent to the landfill. The community is served from a public water supply. Limited barriers exist to prevent local residents or animals from entering the site. There are approximately 10,000 people living within a 3-mile radius of the site. Several small creeks are located near the landfill.

The groundwater and soil are contaminated with heavy metals and chloroform. Touching or accidentally ingesting the contaminated soil could lead to potential health hazards. Pollutants may seep from the landfill into the groundwater. Ingestion of contaminated groundwater may pose a health hazard. This site is being addressed in three stages: immediate actions and two long-term remedial phases focusing on cleanup of the landfill and contaminated groundwater.

Immediate Actions: In 1984, the parties potentially responsible for the contamination posted signs and installed a partial fence around the site; they also covered the landfill and the discarded mercury-containing thermometers with topsoil.

Landfill: The potentially responsible parties began a study in 1984 to evaluate the nature and extent of the contamination associated with the landfill wastes. The work was completed in 1991. In September 1991, EPA selected a remedy that called for installation of a cap on the landfill. The design for the cleanup activities began in late 1992 and is scheduled for completion in 1995.

Groundwater Contamination: The studies conducted to determine the nature and extent of groundwater contamination at the site and to evaluate various cleanup alternatives were completed in the fall of 1993. In October 1993, EPA selected a remedy that allowed for natural attenuation of the contaminants in the groundwater, institutional controls to restrict its use until it reaches safety levels, and groundwater monitoring.

A Consent Order was signed with Becton Dickinson, in which the company was responsible for immediate corrective actions at the landfill in 1984. An Administrative Order also was issued by the EPA in 1984 to Becton Dickinson to study the nature and extent of contamination at the site. A Unilateral Administrative Order was issued by EPA on September 30, 1992 to four industries and three present and past owners requiring them to implement the remedy for the landfill. The immediate actions described above limited the access to the site, reducing the potential for exposure to hazardous materials at the Juncos Landfill site while cleanup activities are being designed.

2. Monroe Township Landfill Site, NJ. Region 2.
EPA ID# NJD 980505671
Date Listed on NPL: September, 1983
Date of Final Remedy Selection: 1993

The Monroe Township Landfill is located on an 86-acre site in Middlesex County, New Jersey. Monroe Township was the original owner and operator of the landfill and continues to own the property. The Township operated the landfill from the mid-1950s until 1968 when it was leased to Princeton Disposal Service for operation under the service contract to the Township. Browning-Ferris Industries of South Jersey (BFISJ) acquired Princeton Disposal Service in 1972 and operated the landfill until 1978. The NJDEPE ordered the site closed in 1978, when leachate outbreaks seeped onto a nearby street.

On October 19, 1979, an Administrative Consent Order (ACO) was signed by BFISJ and NJDEPE establishing methods and schedules for designing and implementing a closure plan. In accordance with the 1979 ACO, the following remedial measures were completed in 1984:

- Installation of a 7,000-foot long compacted clay cut-off wall circumscribing most of the site
- Construction and operation of a leachate collection and storage system which discharges to a Publicly Owned Treatment Works under a New Jersey Pollutant Discharge Elimination System Permit
- Construction of a protective clay cap covering the northern portion of the site and a soil cap covering the remainder of the site

The site was proposed for inclusion on the Superfund National Priorities List by a notice published in the Federal Register (47 FR 58476), on December 30, 1982. On September 8, 1983, the site was formally placed on the NPL by a notice published in the Federal Register (48 FR 40658). On December 29, 1986, BFISJ and the NJDEPE signed another ACO and the following additional remedial measures were completed between 1987 and 1991:

- Upgrading the soil erosion and sediment control systems by replacing former channels with rip-rap lined channels and upgrading the sedimentation basin
- Installation of a seven-foot high chain-link fence surrounding the site to prevent unauthorized access
- Closure of the previous leachate storage lagoon and construction of an underground leachate storage tank
- Installation of an emergency power generator as a contingency for the leachate collection system in case of power failure
- Installation of 13 gas vents for gas ventilation under a New Jersey Air Pollution Control Permit

In accordance with the 1979 and 1986 ACOs, a Remedial Investigation comprising several environmental investigations was performed. The RI developed a conceptual model of the site hydrogeology and assessed the nature and extent of contamination in various environmental media including groundwater, surface water, surface soil, stream sediments and landfill gas. All samples were analyzed for Target Compound List/Target Analyte List compounds. The analytical results indicated no significant levels of contaminants in soil, surface water and sediments. Contaminants detected in the gas vents were within the permit limits. Some contaminants were detected in on-site groundwater (within the containment wall) including arsenic, cadmium, lead, nickel, benzene, chlorobenzene, 1,2-dichloroethane, 1,1-dichloroethene and vinyl chloride. Arsenic levels were attributed to natural background conditions. Although low levels of contaminants were detected in one on-site monitoring well located outside the containment wall, no discernible contaminant plume was found. In addition, no contaminants were detected in the off-site monitoring wells installed downgradient of this location. Furthermore, groundwater modeling has indicated that natural attenuation will prevent the off-site migration of contaminants which are above levels of concern. A Risk Assessment (RA) was performed based on the results of the RI. The RA concluded that the Site poses no current or future unacceptable risk to public health and the environment.

Based on the results of RI and the RA, it appears that the source control measures undertaken by BFISJ are effective in controlling any off-site migration of contaminants. Therefore, on April 23, 1993, NJDEPE signed a Record of Decision for this site, selecting "No Further Action" to address this site. The ROD also calls for the implementation of a groundwater monitoring program and maintenance of the existing source control measures system. BFISJ will be required to begin monitoring selected on-site and off-site groundwater monitoring wells within 6 months of signing the ROD. This monitoring will be conducted on a quarterly basis for the first 5 years and thereafter on a yearly basis.

Because this remedy will result in hazardous substances remaining at the site, a review will be conducted within 5 years of signing the ROD to ensure that the remedy continues to provide adequate protection of human health and the environment. Having met the deletion criteria, EPA proposes to delete this site from the NPL. EPA and NJDEPE have determined that the response actions conducted to date are protective of human health and the environment.

3. Ringwood Mines/Landfill Site, NJ. Region 2.
EPA ID# NJD980529739
Date Listed on NPL: September, 1993
Date of Final Remedy Selection: July, 1993

The Site consists of approximately 500 acres in a historic mining district in the Borough of Ringwood, which is located in the northeast corner of Passaic County, New Jersey. The Ringwood Mines are a series of iron ore mines that operated almost continuously from the mid-1700s to the early 1900s. In 1965, Ringwood Realty Company, a wholly-owned subsidiary of the Ford Motor Company, obtained control of the Site property. Beginning in 1967 and until the mid-1970s, Ringwood Realty used the Site to deposit waste products from the Ford factory in Mahwah, New Jersey. The waste products included car parts, solvents and paint sludge. Some of these wastes were deposited on the ground surface in natural and man-made depressions, and some were allegedly dumped into the mine shafts.

Pursuant to a March 1984 Section 3013 RCRA Administrative Order on Consent between EPA and Ford, Woodward-Clyde Consultants was retained to perform the field studies and conduct a Remedial Investigation. The RI was conducted in four phases between March 1984 and April 1988 under EPA oversight. Six different media were sampled during the RI: seep water, soils (paint sludge waste), overburden (upper aquifer) groundwater, deep bedrock groundwater, surface water and stream sediments. Results were as follows:

A groundwater contaminant plume was not identified for any of the contaminants found in any areas at the Site. Groundwater contamination occurred at a low level, and was scattered and generally confined to paint-sludge locations. No detectable migration of groundwater contamination was identified. Three rounds of surface

Appendix 3 — Excerpts From NPL Site RODs

water samples were collected along with seep-water samples. No significant contamination was found. Arsenic was found in stream sediment samples from Park Brook and Peters Mine Brook. The highest concentration found was 31 ppm. Arsenic concentrations as high as 13.5 ppm were found in upstream samples. Paint-sludge waste was identified at four locations at the Site. The paint sludge was sampled and analyzed to determine a waste classification. The paint sludge was identified as EP (extraction procedure) toxic for lead.

Beginning in October of 1987, Ford and its contractors (in accordance with an EPA approved work plan) excavated and removed 7,000 cubic yards of surficial paint sludge containing lead and arsenic from four areas at the Site under an Administrative Order issued by EPA in June of 1987. The paint sludge was disposed of at an out-of-state facility in compliance with State and Federal regulations. The Record of Decision for the Site was signed on September 29, 1988. EPA's selected remedy for this Site had three components:

1. Achieving health-based levels, including State and Federal MCLs, in the upper aquifer of the Site through natural attenuation processes. Remediation of groundwater contamination was evaluated and rejected in the ROD since extraction of the groundwater would have diluted levels to below treatment standards. The low levels and sporadic occurrence of groundwater contamination make groundwater treatment impractical. In addition, since the paint sludge removal has eliminated the suspected source of surficial groundwater contamination, groundwater quality should improve without further remediation.
2. Implementing a long-term surface water and groundwater monitoring program to confirm that groundwater contamination meets or is below health-based levels and to protect against future threats to the groundwater and surface water throughout the Site.
3. Performing confirmatory test pitting and soil sampling, along with possible removal of contaminated soils or sludge.

In October of 1989, additional paint sludge was uncovered in the southern section of the O'Connor Disposal Area within the Site. During the excavation of this additional paint sludge which began in January of 1990, a total of 61 drums were discovered, some which contained liquid and solid waste. Approximately 20 (55-gallon) drums of liquid and solid waste were removed and disposed of off the Site. 17 one cubic yard pelletized containers which contained excavated drums and their contents, three drums containing residual materials associated with the Site, and 727 tons of additional paint sludge were also disposed of off the Site. Further geophysical surveys and test pit work were conducted in the O'Connor Disposal Area in 1992 and 1993. No further barrels of hazardous substances were discovered.

Ford, under EPA oversight, has been implementing the Environmental Monitoring Program (EMP) which is a five-year program that is being conducted pursuant to an EPA Administrative Order on Consent executed on August 29, 1989. After the five-year EMP, EPA will re-evaluate the monitoring results to ensure that groundwater continues to pose no further threat to human health or the environment. Dependent upon this re-evaluation, long-term monitoring of the groundwater may continue for a period of up to 30 years. Presently, the shallow aquifer is not being used as a potable water source. State restrictions on shallow wells should remain in effect for the foreseeable future.

All the requirements for the deletion of this Site from the NPL have been met. Post-excavation confirmatory sampling has verified that all removal action criteria for the removal of paint sludge were met. Extensive geophysical studies along with exploratory test pitting operations did not uncover any further contaminated soils/sludge, or barrels of hazardous substances at the Site. A conservative assessment of risk attributable to the release of hazardous substances from the Site indicated that the current risk posed by the Site is within an acceptable range. A long-term monitoring program has been implemented which provides further assurance that the Site no longer poses any threats to human health or the environment. The State of New Jersey, in its letter of July 23, 1993, concurred on the deletion of this Site from the NPL.

1994

1. Coakley Landfill, NH. Region 1.
EPA ID# NHD064424153
Date Listed on the NPL: June, 1986
Date of Final Remedy Selection: Fall, 1994

The privately owned Coakley Landfill site is a 92-acre parcel of land that was operated by several municipalities. The landfill area encompasses 27 acres in the southern portion of the site. The site accepted municipal and industrial wastes from the Portsmouth area between 1972 and 1982 and incinerator residue from the incineration recovery plant for the Refuse-to-Energy Project between 1982 and 1985. VOCs and metals are the predominant contaminants found. On- and off-site surface water and groundwater are contaminated. The site is located on a groundwater/surface water divide, and residential wells to the south, southeast, and northeast of the landfill are contaminated with low levels of VOCs. Public water service has been extended to the areas with contaminated wells by local communities. There are several small commercial facilities, motels, and restaurants nearby.

The site is being addressed through Federal, State and potentially responsible parties' actions. On-site groundwater is contaminated with arsenic, phenol, and methyl ethyl ketones; off-site groundwater is contaminated with heavy metals including arsenic, chromium, and lead, and VOCs including benzene and methyl ethyl ketones. On-site soil sediments are contaminated with arsenic and lead. Stream sediments contain contamination from arsenic and VOCs. Leachate contamination at the site includes VOCs, tetrahydrofuran, and ketones. Metals and VOCs were detected in nearby wetlands. Potential use of groundwater as a water supply is the main threat to human health.

The site is being addressed in three stages: initial actions and two long-term remedial phases focusing on source control and cleanup of off-site groundwater. *Initial Actions:* In 1989, North Hampton extended a municipal water line to residents who had been supplied by 13 private wells contaminated with VOCs. The State set up a residential well-monitoring system with an early warning system to detect any groundwater contamination in the area. Most area residents now have access to municipal water or uncontaminated groundwater.

Source Control: The State conducted an investigation from 1986 to 1987. The goals of the field work were to characterize the hydrogeologic conditions at the site, including an estimate of the total area of the landfill and soil deposits, details of the hydraulic properties of bedrock and selected surface streams, and the identity of pathways for contaminant migration from the site.

The State completed the study in 1990. Based on the results of the study, a cleanup remedy was selected which includes consolidating approximately 2,000 cubic yards of wetland sediments; consolidating approximately 30,000 cubic yards of on-site solid waste; fencing and capping the landfill; collecting and treating landfill gases by thermal destruction; extracting groundwater and treating it with a combination of chemical, biological, and physical processes; and establishing long-term monitoring and institutional controls. Design of the remedy began in the summer of 1992, and is scheduled for completion in 1995. The EPA began a study in 1990 of the migration of contaminants into off-site groundwater and the ecological effects of the site contamination on adjacent wetlands. In the fall of 1994, a groundwater remedy was selected which includes imposing institutional controls to prevent the use of contaminated groundwater, allowing the groundwater to clean itself through natural attenuation, and long-term monitoring. Final cleanup actions are expected to begin in the fall of 1995.

The State issued a Consent Order in 1983 requiring the owner to accept only incinerator ash from the Refuse-to-Energy Project. As of February 1990, notices had been sent to 60 parties potentially responsible for the site contamination. The provision of an alternate drinking water source has reduced the potential for exposure to contamination, making the Coakley Landfill safer while further cleanup activity is being planned.

2. Rockingham Landfill, VT, Region 1.
EPA ID# VTD980520092
Date Listed on the NPL: October, 1989
Date of Final Remedy Selection: Fall, 1994

In the early 1960s, the 17-acre Browning-Ferris Industries, Inc. (BFI) Sanitary Landfill (Rockingham) site served as a borrow area for the construction of Interstate 91. In 1973, the site was bought from an individual who had started operations in 1968. State files indicate that industrial wastes, including heavy metals, bases, pesticides, and volatile organic compounds (VOCs) were deposited in the unlined disposal area from 1968 to 1979. In 1983, Vermont licensed the site as a municipal landfill certified to accept hazardous waste from small-quantity generators. The landfill was closed in 1991. The Vermont Department of Environmental Conservation (VT DEC) reports that nearby residential and monitoring wells downgradient of the site have been contaminated since 1979. The contaminated residential wells are no longer in use. There are two leachate collection ponds on site. A tar cap covers a portion of the landfill to prevent the infiltration of rainwater; however, cracks in the cap have been observed, and it is covered with new refuse. Approximately 2,700 people live within a mile of the site, and 6,400 residents live within 3 miles. Three homes near the site are supplied water from a water supply line provided by BFI. The Connecticut River is located 560 feet to the east, along the drainage route of surface water leaving the site.

The site is being addressed through Federal and potentially responsible parties' actions. The soil and groundwater contain contamination from VOCs and heavy metals including chromium, copper, lead. Drinking water from contaminated wells in the area poses a threat to public health. The Connecticut River also may receive contaminants from groundwater discharge, posing a threat to water quality and aquatic life.

The site is being addressed in two stages: initial actions and a long-term remedial phase directed at cleanup of the entire site. BFI is providing an alternate drinking water supply to residences with contaminated wells. In 1989, BFI installed an active gas collection system to control landfill gases. In early 1993, BFI also installed a groundwater interceptor trench to collect surface water and leachate seeps that were discharging into the Connecticut River. In 1993, the EPA completed studies at the site and initiated an early action to address the source of contamination under the Superfund Accelerated Cleanup Method (SACM). In 1994, the potentially responsible parties capped the landfill and expanded the existing active gas and leachate collection system.

In the fall of 1994, a remedy was selected to address the cleanup of the groundwater. Through management of existing site controls, including the landfill cap, active gas and leachate collection systems, and long-term monitoring, the groundwater will clean itself naturally over time. Long-term monitoring of the groundwater, surface water, and residential wells will continue to ensure that established cleanup levels are met.

The State issued three orders to the owner between 1980 and 1983, requiring BFI to determine the hydrogeology of the landfill, monitor on-site groundwater, and provide drinking water to affected residents nearby. In July 1992, the EPA entered into an Administrative Order on Consent with two parties, BFI, Inc. and Disposal Specialists, Inc., to perform the site investigations. An Administrative Order by Consent was signed between the EPA and the potentially responsible parties, BFI, Inc. and Disposal Specialists, Inc., to perform the initial actions.

Installation of the landfill the cap and gas and leachate collection systems has reduced the potential for exposure to contaminated soil and groundwater, making

Appendix 3 — Excerpts From NPL Site RODs

the BFI Sanitary Landfill site safer while monitoring and natural restoration of groundwater continues.

3. Davie Landfill, FL, Region 4.
EPA ID# FLD980602288
Date Listed on the NPL: September, 1983
Dated of Final Remedy Selection: 1994

The Davie Landfill site, consisting of an 80-acre trash landfill, a 30-acre sanitary landfill, and a 10-acre sludge lagoon near the intersection of Orange Drive and Boy Scout Road, began operation in 1964, accepting trash and ash from the County's adjacent garbage incinerator. The sludge lagoon was constructed in 1971 in an unlined natural depression on site to accept grease trap pump-outs and septic tank and treated municipal sludge. The lagoon overflowed on several occasions, resulting in surface water discharges to an adjacent borrow pit. The sludge lagoon was closed in 1981.

The incinerator was closed in 1975 because the excessive particulate emissions failed to meet new air regulations. The sanitary landfill was opened to replace the closed incinerator. Landfilling activities ceased in 1987, when the facility reached its design capacity. The solid waste landfill was used to dispose of the municipal solid waste being burned at the on-site incinerator. Construction debris, tires, and other wastes that could not be incinerated also were placed in the solid waste landfill. Dairy farms, ranches, and horse stables are located in the vicinity of the site. Approximately 50 homes are located to the south of the site; the nearest residence is ½ mile away. There are 5 wells within 500 feet of the site and 21 within ¼ mile. All municipal water supplies in the area receive water drawn from the Biscayne Aquifer. The aquifer is the sole source of potable water for about 10,000 residents in the area.

The groundwater and the water in the borrow pits on site and downgradient of the site show elevated levels of sulfate, chloride, lead, and ammonia. Antimony, benzene, vinyl chloride, and other contaminants have been detected in monitoring wells and private wells south of the landfill. Sludge from the lagoon was found to contain cyanide and sulfides. Potential health threats include accidental ingestion, inhalation, and direct contact with contaminated soil, groundwater, surface water, and sediments. The site is fenced, and access to the site is restricted. The site is being addressed in two long-term remedial phases focusing on cleanup of the sludge lagoon and groundwater.

Cleanup technologies chosen to address sludge lagoon contamination included: dewatering and stabilizing the sludge lagoon contents; placing treated sludge lagoon contents in a lined sanitary landfill cell; and installing an approved cover on the cell. The State required the County to provide service connections to the municipal water supply system for each affected residence near the site. The County offered affected residents bottled water until the water lines were functional. The alternate water supply is in place. The County initiated site construction on the sludge lagoon in 1989, and cleanup activities are completed.

An investigation into the nature and extent of groundwater contamination at the site began in early 1992 and was completed in 1994. Vinyl chloride and antimony concentrations in groundwater were detected. Due to the low levels of contamination detected, natural attenuation was selected as the remedy. In addition, EPA required the County to monitor residential areas and provide service connections to the municipal water supply system for affected residences near the site. Designs for the municipal water supply connections are currently underway and are scheduled for completion in mid-1995. The provision of an alternate water supply and the completion of the sludge lagoon cleanup activities have reduced the danger of exposure to contamination while additional municipal water supply lines are being designed.

4. B & B Chemical Company, Inc. FL, Region 4.
EPA ID# FLD004574190
Date Listed on the NPL: August, 1990
Date of Final Remedy Selection: 1994

The B & B Chemical Company, Inc. has manufactured industrial cleaning compounds on this 2-acre site in Hialeah since 1962. The company prepares its products in mixing vats, which, along with the company's tank trucks, are washed down once a year. Before 1976, the wash water was put into unlined lagoons. Since then, the company has run it through a treatment system before discharging it to the Hialeah sewers. Officials have been concerned about the impact of the lagoons on groundwater quality since 1975. The underlying Biscayne Aquifer supplies drinking water for all of Dade County. This site is in a highly industrialized area. Four public well fields are within 3 miles of the site and serve approximately 750,000 people. One well is within 3,000 feet of the site. Production from the well fields has been curtailed due to groundwater contamination. The Miami Canal is 800 feet to the southwest of the site.

In 1985, the EPA found VOCs including chlorobenzenes and dichloroethylene from former manufacturing operations in monitoring wells on and off the site; they also found chromium in on-site wells. Drinking or coming into direct contact with polluted groundwater may pose a health threat.

B & B Chemical Company, under an agreement with Dade County, operated the groundwater recovery and treatment system at the site until July 1989, when they unilaterally stopped recovery of the groundwater. Groundwater recovery was restarted in November 1989. In September 1994, EPA chose a remedy which includes the natural attenuation of groundwater contaminants, imposition of institutional controls over the south central part of the property, and groundwater monitoring to verify that cleanup through natural attenuation is occurring. Given the currently low contaminant concentrations in the groundwater and the observed decreasing trends in concentrations, it is anticipated that natural attenuation will further reduce groundwater contamination to established cleanup standards within 2 years. The earlier groundwater treatment performed by the potentially responsible party has reduced the potential for exposure to contaminants from the B & B Chemical Company, Inc. site while natural attenuation of the contaminated groundwater is taking place.

1995

1. Wilson Concepts Site, Pompano Beach, FL, Region 4.
EPA ID# FLD041184383
Date Listed on the NPL: March, 1989
Date of Final Remedy Selection: 1992

The Wilson Concepts Site occupies approximately two acres in a highly industrialized section of northeastern Broward County, Florida. Associated operations at the facility included precision machining, drilling, and milling of metal parts, vibratory deburring, degreasing, steam cleaning, and spray coating of parts. A variety of chemicals were used, including organic solvents, chlorinated solvents, petroleum products, paints, cyanides, acids, and bases. From 1976 through 1989, several inspections were conducted by Broward County Environmental Quality Control Board (BCEQCB) which documented poor waste-handling practices, including discharge of industrial wastes onto the ground. Raw materials usage at the Site was documented on two occasions over a period of 10 years. In the early 1980s, possibly as early as 1981, Wilson Concepts submitted a hazardous materials inventory list to the BCEQCB. The chemicals used at the Site included a variety of hydraulic and lubricating oils, metal protection agents, water coolants, methylene chloride, methyl ethyl ketone, and chemical cleaners (possibly corrosives). In August 1985, EPA conducted a Preliminary Assessment of the Site and in July 1986 requested its contractor, NUS, to perform a Sampling Investigation. The results of this sampling caused the Site to be proposed for the NPL in July 1988. In March 1989, the Wilson Concepts Site, was formally included on the NPL.

The Remedial Investigation was conducted in two phases, phase I in 1990 and phase II in 1991. The groundwater contaminant levels detected at the Site during the RI were much lower than those detected during the Site Listing Investigation (SLI). Most of the contamination consisted of organic compounds which easily evaporate and/or naturally degrade. Trichloroethylene was detected in two monitoring wells during the first sampling event but was not detected above groundwater standards in phase II, indicating that no distinct groundwater contaminant plume existed. Chromium and strontium along with low levels of organics including toluene were detected in Site soils. Chromium was observed in two locations in subsurface soils at elevated levels. Though similar organic contaminants were detected during the RI, the concentrations were much lower than previously detected and were within the range considered by EPA to be safe for human exposure. It is possible that the organic contamination was reduced by natural attenuation.

The Record of Decision, issued by EPA, Region 4, on September 22, 1992 selected alternatives consistent with the recommendation in the Feasibility Study. The remedy is a ``No Action'' remedy which included monitoring of on-site groundwater, quarterly, for a one-year period to confirm the appropriateness of the no action remedy. A Post ROD Project Operations Plan for the Site was completed on May 25, 1993. Field work for Post ROD groundwater monitoring begin on June 21, 1993, and was completed on July 18, 1994. Post ROD groundwater monitoring confirmed that Site related contaminants had declined in concentration and are below maximum contaminant levels (MCLs). Sporadic detections of iron, manganese and aluminum do not appear site related. During the Remedial Investigation drums were left on the Site. Those drums since have been sampled. No contaminants of concern were detected in either the drummed soil or aqueous samples. The Agency for Toxic Substances and Disease Registry (ATSDR) completed a health assessment for the Site in May, 1990. ATSDR reviewed the surface water and groundwater data and recommended that the Site should not be considered for follow-up health studies due to no present exposure to population.

No institutional controls are necessary for the Site. A five-year groundwater review will not be conducted because contaminants of concerns have achieved levels below MCLs, which was verified through Post ROD monitoring. These levels indicated that no future threats to the public health or environment exist. EPA and the State find that the remedy continues to provide adequate protection of human health and the environment. Therefore, EPA proposes to delete the Site from the NPL. EPA, with concurrence of the State of Florida, has determined that all appropriate Fund-financed responses under CERCLA at the Wilson Concepts Site have been completed, and that no further cleanup by responsible parties is appropriate.

2. Diamond Shamrock Site, GA, Region 4.
EPA ID# GAD990741092
Date Listed on the NPL: 1990
Date of Final Remedy Selection: 1994

The 1-acre Diamond Shamrock Corp. Landfill site is located north of West Girard Avenue and is adjacent to, and east of Cedar Creek in Cedartown, Georgia. In 1972, the company buried drummed and bulk waste in five trenches approximately 6-foot deep at the landfill. According to the company, the waste included fungicides, amides, oil, oil sludges, esters, alcohols, and metallic salts. The unlined trenches are located in an area of permeable soils within the flood plain of Cedar Creek, which is a major tributary of the Coosa River. Area groundwater underlying the site is shallow. The current water supply for the City of Cedartown is a spring. Cedar Creek has been used for fishing and possibly for swimming.

On-site groundwater and surface water are contaminated with manganese, toluene, trichloroethene, and 1,2-Dichloroethane from wastes deposited on the site. Soil contained these same contaminants prior to cleanup. Potential health threats include direct contact with or accidental ingestion of contaminated groundwater.

The site is being addressed in two stages: initial actions and a long-term remedial phase focusing on cleanup of the groundwater.

In 1990, the EPA recovered and removed an estimated 1,800 cubic yards of soil, solid waste, and debris (empty drums), which were shipped off site for disposal in an EPA-approved industrial landfill. Approximately 680 drums were removed. An estimated 1,500 cubic yards of waste-impacted soil was prepared on site and then placed into two active aeration biotreatment cells built east of the site. Data from the soil samples collected from the treatment cells indicates that the bioremediation techniques are effectively reducing the concentrations of the appropriate contaminants. Approximately 8,400 gallons of liquid waste were recovered during excavation and incinerated at a licensed hazardous waste disposal facility.

Operating under EPA oversight, Henkel Corporation completed an investigation into the nature and extent of contamination at the site in 1994. The investigation concluded that the surface water, sediments, and soils posed no threat after initial cleanup actions were completed, but that groundwater contamination still needed to be addressed. The EPA selected a remedy in 1994 that includes restricting groundwater usage and monitoring groundwater and surface water to confirm that they are cleaning themselves through natural attenuation.

The removal of contaminated drums and liquid waste and the treatment of contaminated soil have reduced the threat of direct exposure to pollutants by the surrounding community and the environment while the groundwater and surface water are cleaned through natural processes. The groundwater will be monitored to ensure that contamination levels continue to diminish.

Site access restrictions, and groundwater and surface water monitoring will be used to ensure that natural attenuation will be effective to prevent migration of contaminants.

3. Jackson Township Landfill Site, NJ, Region 2.
EPA ID# NJD980505283
Date Listed on the NPL: September, 1983
Date of Final Remedy Selection: September, 1994

The Jackson Township Landfill Site is located off Lakehurst Avenue in Jackson Township, Ocean County, New Jersey. Jackson Township purchased the 135-acre landfill, which is situated in a regional reserve known as the New Jersey Pinelands, in 1972. The property was previously owned and mined by Glidden Corporation. Approximately 20 acres of the property were used for the disposal of various wastes.

In 1977, there were multiple complaints of medical problems associated with the use of area groundwater. Subsequently, the NJDEP ordered groundwater analyses to be conducted in the vicinity of the landfill. Based upon the results of these analyses, NJDEP concluded that a segment of the Cohansey aquifer and several domestic wells had been contaminated by hazardous substances disposed of at the Jackson Township Landfill. The NJDEP used Spill Fund monies to provide bottled potable water for residences impacted by the groundwater contamination.

In 1978, NJDEP ordered Jackson Township to stop disposing of liquid wastes at the landfill. In 1980, a citizen lawsuit resulted in a municipal water system extension to properties affected or potentially affected by contaminants disposed of at the landfill. The landfill was closed by order of the Superior Court of New Jersey in February 1980.

Sampling of 22 monitoring wells and eight domestic wells was performed in December 1981 and February 1982. Results of this sampling indicated that contaminants were present in groundwater at levels only slightly exceeding criteria established for the protection of groundwater. In April and December of 1982, the NJDEP sampled seventeen shallow and deep monitoring wells at and in the vicinity of the landfill. Organic compounds were only detected above method detection limits in one well, and inorganics rarely exceeded established criteria during this sampling event. Additional groundwater sampling conducted in 1985 revealed similar results.

In December 1982, the Jackson Township Landfill was included on the National Priorities List of Superfund sites. In 1988, the NJDEP and Jackson Township reached an agreement, known as the Judicial Consent Order (JCO), which required Jackson Township to reimburse the NJDEP for Spill Compensation Fund monies spent by the NJDEP. In addition, the JCO required Jackson Township to arrange for and fund the investigation and remediation of the landfill. Throughout 1989 and 1990, a Remedial Investigation was conducted in which air, surface water, groundwater and soil studies were performed.

During performance of the RI, 22 groundwater monitoring wells at and around the landfill were sampled. In addition, four surface water and sediment samples were collected in the Ridgeway and Obhanan-Ridgeway Branches, which are both tributaries to the Toms River. No site-related compounds were detected in surface water samples above Federal or State Surface Water Quality Criteria. Similarly, no compounds were detected in sediments above levels of concern developed in the Risk Assessment for the site.

Nine soil borings were also installed at the Jackson Township Landfill during performance of the RI. Contaminants were only detected in soils at low levels. Results of the RI indicate that contaminant levels have continued to decrease due to natural attenuation. A Risk Assessment was conducted based upon the results of the RI. The Risk Assessment concluded that there is no unacceptable current or potential future risk to public health and the environment associated with the landfill. The September 27, 1994 Record of Decision for the Jackson Township Landfill site selected the No Further Action remedy, because the Risk Assessment has shown that no further action is necessary to protect human health and the environment. Furthermore, the ROD provides for the performance of a review within five (5) years of signing the ROD to ensure that the no-further-action remedy continues to provide adequate protection of human health and the environment.

1996

1. New Castle Spill Site, DE, Region 3.
EPA ID# DED058980442
Date Listed on the NPL: September, 1983
Date of Final Remedy Selection: September, 1996

The Site is a former manufacturing plant of the Witco Corporation located 0.5 miles west of the Delaware River and 0.5 miles north of the City of New Castle, Delaware. Surrounding the Site is a mixed commercial and residential area. The Site is bordered on the west by a marsh and on the east by a dual highway. Among the chemicals Witco used in the production of plastic foams was the semi-volatile organic compound tris(2-chloropropyl) phosphate (Tris). Sometime before 1977 it is estimated that approximately 4–5 drums of Tris, stored on the Site, were spilled on the ground, contaminating the soil and shallow groundwater beneath. Under the direction of DNREC, the groundwater was pumped and discharged into the adjacent marsh. Numerous investigations of the soil and groundwater followed, including an EPA Site Inspection in 1981. EPA proposed the Site for inclusion on the NPL on December 30, 1982 and finalized the listing on the NPL on September 8, 1983.

Pursuant to an Administrative Order on Consent with DNREC, Witco conducted a remedial investigation (RI) and feasibility study (FS) from February 1988 to June 1989. These studies determined the extent of contamination, the risks to human health and the environment posed by the contamination, and cleanup alternatives to address those risks. The RI included sampling of soils, groundwater, surface water, and marsh sediments. Results of the RI showed that the groundwater in the shallow Columbia aquifer was contaminated with the organic compounds Tris, TCE, and 1,2-dichlorethene. Only Tris was determined to be present at levels that presented a significant risk to human health. TCE was determined to be from another source upgradient of the Site and was addressed through a separate State action. No Tris contamination was found in the deeper Potomac aquifer. Tris and several other organic compounds were found in soil samples but at levels that would not threaten human health or the environment and were no longer considered a source of contamination to the groundwater. Contaminant levels found in the marsh area were well below levels that would threaten the wetland habitat or environmental receptors.

Using the RI data, an endangerment assessment was performed to evaluate the risks that contaminants detected at the Site posed to human health and the environment. Of the numerous exposure pathways evaluated, only potential future exposure to groundwater used as a potable water supply was determined to present a risk to human health that exceeded acceptable levels as defined by the NCP. As no one was using the Columbia aquifer in the area for a potable water supply, natural attenuation was determined to be the most appropriate means by which to reduce the Tris concentrations to acceptable levels. EPA developed a health-based drinking water cleanup level of 4.4 mg/l for Tris and estimated that it would take approximately four years for Tris to reach this level by natural attenuation.

To document this cleanup approach, EPA and DNREC issued a Record of Decision on September 28, 1989 which included the following components: (1) monitoring of the Columbia aquifer on a quarterly basis for Tris to ensure the effectiveness of the natural attenuation process; (2) monitoring of the Potomac aquifer on an annual basis for Tris to ensure that contamination has not migrated from the Columbia aquifer; (3) monitoring of the surface water and sediments of the adjacent wetlands on an annual basis for Tris, with further evaluation and bioassay testing required if trigger values of 100 µg/l Tris in surface water, or 1000 µg/kg Tris in sediments were reached; (4) institutional restrictions on the placement of wells in the Columbia aquifer in the vicinity of the Site; and, (5) a 5-year effectiveness review of the remedy.

In April 1991, EPA and Witco entered into a Consent Decree whereby Witco agreed to implement the remedy selected in the ROD. Witco began quarterly groundwater, surface water, and sediment monitoring in July 1992, which continued through September 1995. Tris levels in the surface water and sediment samples were consistently well below the trigger levels specified in the ROD or not detected at all; therefore, no further evaluation or bioassay testing was necessary in the marsh. Tris was not detected in the Potomac aquifer in any sampling event.

Of the 13 monitoring wells screened in the Columbia aquifer that were included in the monitoring program, only two wells showed concentrations of Tris above the groundwater cleanup level of 4.4 mg/l during the entire monitoring period. By natural attenuation, Tris concentrations decreased with time in these two wells until they were below the cleanup level for the last several sampling events. During the last sampling event in September 1995, Tris concentrations ranged from approximately 1 to 2 mg/l. A statistical analysis of the data confirmed that there is very little chance that the Tris concentration will exceed the cleanup level in the future.

In November 1990, pursuant to the ROD, DNREC instituted a Ground Water Management Zone (GMZ) in the vicinity of the Site to restrict installation of drinking water wells in the area. Now that the Tris cleanup level has been achieved in the area of the Site and there is no longer a need to prevent exposure to the groundwater, DNREC will retract the GMZ following the deletion of the Site from the NPL. Based on the information presented above, EPA has determined that Witco, the responsible party for this Site, has implemented all response actions required and that no further action is appropriate. Thus, the required NPL deletion criteria presented in Section II, above, have been met. DNREC has concurred on this determination.

The ROD stated that EPA would conduct a five-year effectiveness review to reevaluate the Site. The evaluation made to determine if the NPL Deletion criteria have been met serves as that review. In addition, EPA reviewed the most recent toxicological information available for Tris and determined that the cleanup level of 4.4 mg/l in groundwater remains protective. Therefore, EPA has determined that the Site poses no significant threat to public health or the environment.

The NCP at 40 CFR 300.430 states that EPA shall review remedial actions every five years if hazardous substances, pollutants, or contaminants remain at the site above levels that allow unrestricted exposure and unlimited use. Since neither of these conditions exists at this Site, further five-year reviews are not warranted and will not be conducted. EPA, with the concurrence of DNREC, believes that the criteria for deletion of the Site have been met. Therefore, EPA is proposing deletion of the Site from the NPL.

2. Alcoa (Vancouver Smelter) Site, WA, Region 10.
EPA ID# WAD009045279
Date Listed on the NPL: February, 1990
Date of Final Remedy Selection: 1996

The Alcoa Site is located in Vancouver, Clark County, Washington, approximately 3 miles northwest of downtown Vancouver, Washington and approximately 300 to 500 feet north of the Columbia River. The site is found at the southeastern corner of the VANALCO smelter complex located at 5701 NW Lower River Road, Vancouver. The site has been used for industrial purposes since World War II and is currently zoned for heavy industry. The area is changing from a mixture of agriculture and heavy industry to commercial and heavy industry. The site consists of three waste piles, contaminated soil under waste piles and subsurface contaminated strata and groundwater.

The Alcoa facility has produced aluminum since 1940 using the Hall-Heroult electrolytic cell process. The process is an electrochemical reduction reaction in which aluminum oxide is dissolved in a bath of molten salts (cryolite) at a temperature of 1760 degrees. An electric current is passed through the cell causing the reduction of alumina to aluminum. The entire process occurs in a steel shell or pot that is lined with insulation material and carbon, known as potlining.

In order to retain the purity of the molten aluminum and the structural integrity of the cell, the molten aluminum and cryolite mixture must be kept isolated from the steel shell of the pot. Over time, the carbon lining materials become impregnated with the molten cryolite solution, eventually threatening the integrity of the steel and carbon shell. The pot is drained and the carbon and insulation materials replaced. The carbon that is removed from failed pots, is known as spent potlining. Spent potlining consists of carbon, fluoride, aluminuma and sodium, with minor amounts of calcium, silica, iron and cyanide and is a listed (K088) dangerous waste.

- Early 1950s-1973: Spent pot lining was shipped off-site to the Reynolds Aluminum Plant in Livingston, Washington and recycled.
- 1973 to 1981: Pot liner waste piles were formed on site. They were not covered and were exposed to normal precipitation. Fluoride and cyanide leached out of the exposed pot liner and contaminated soils and groundwater below the piles.
- 1977: Alcoa installed nine shallow monitoring wells in the vicinity of the three waste piles. Sampling of these wells discovered groundwater contamination.
- 1978, 1981: The piles were covered with plastic and clean sand.
- 1986: As a result of increasing cyanide in the monitoring wells, Washington Department of Ecology ordered Alcoa to conduct a program to assess the groundwater contamination at the site and to evaluate potential cleanup actions.
- 1987: Alcoa submitted to Washington Dept. of Ecology a Remedial Investigation and Feasibility Report. The investigation revealed that the groundwater contamination extended to the Columbia River. The report identified four water-bearing zones at the site, three of which were contaminated with concentrations of cyanide and fluoride above drinking-water standards.
- 1989: EPA identified Washington Deptartment of Ecology as the Lead Agency for cleanup activities at the site. The Agency for Toxic Substances and Disease Registry conducted a site visit, reviewed available data and made several recommendations regarding remediation.

- 1990: The site was placed on the NPL by EPA.
- 1992: Washington Department of Ecology issued a final Cleanup Action Plan (CAP) under MTCA and filed a Consent Decree with Alcoa in State Court. Remedial action was started and completed. Alcoa's final remedial action report was submitted to Washington Department of Ecology.
- 1994: Remedial requirements of the Consent Decree (described in the next section) have been met by Alcoa. Washington Department of Ecology certifies that the construction phase has been completed.
- 1996: Washington Department of Ecology issues a Preliminary Close Out Report (PCOR) and certifies that all remedial action specified in the CAP has been completed, no further action is expected, and that the remedy is protective of human health and the environment.

On February 7, 1992, Washington Department of Ecology, as the Lead Agency and pursuant to MTCA, issued the CAP (equivalent to the CERCLA Record of Decision) for the Alcoa Site. The CAP lists the cleanup goals for the site, presents the different cleanup alternatives that were examined, and presents Washington Department of Ecology's selected site cleanup method.

Source control was accomplished by the removal of 71,758.91 tons of potliner material to Chem-Security Systems, Inc., Arlington, Oregon, a permitted hazardous waste landfill. The potliner material was excavated by using conventional excavation equipment. The contaminated soils beneath the piles were sampled for cyanide and fluoride once the potliner was removed. Capping contaminated soils with 50 mil HDPE or 40 mil PVC liner and covering with 2 feet of sand with top soil. The capped area was fenced and graded to drain. A 50 mil high-density polyethylene (HDPE) flexible membrane liner was placed on compacted sand. The liner extended beyond the limits of the removed pile. A 1 ft. × 1 ft. anchor trench was excavated around the perimeter of the cap to hold the liner in place. 18 to 24 inches of clean sand was placed over the entire area. The sand was placed so the capped area would drain from north to south. Upon completion of the sand cover, 6 inches of topsoil was placed and compacted over the capped area. The topsoil was hydroseeded and the capped area was fenced with an 8-ft. chain-link fence. The purpose of this cap is to minimize further infiltration of water into the contaminated soil and thereby minimize or prevent further leaching of the contamination from the soil into the groundwater. Alcoa has inspected and performed maintenance on the cap on a quarterly basis during the regularly scheduled groundwater monitoring activities. Maintenance requirements for the cap include grading to maintain proper site drainage, repair of any erosion or areas of distressed vegetation, and maintenance of site perimeter fencing and warning signs.

Alcoa has recorded a restrictive land use covenant in the property deed for the site to ensure that no groundwater is removed for domestic purposes from the plume and that there is no interference with the cleanup action. Alcoa may use the site for industrial purposes consistent with the remedial action and the covenant. If the levels of fluoride in the groundwater reach 4.0 mg/l and free cyanide in groundwater reaches 0.2 mg/l, levels that are safe for drinking, Alcoa or the subsequent owner may request that Washington Dept. of Ecology remove the requirement for a restrictive covenant. However, Washington Dept. of Ecology may agree with that request only after a public comment period and insofar as the request is consistent with applicable law, including cleanup standards.

Groundwater remediation will be required if fluoride and cyanide concentrations increase near the Columbia River. The concentration of cyanide and fluoride will have to increase to levels that are treatable. Subsurface flow into the Columbia River is from the deep and aquifer zones. Measurements in the Columbia River upstream and downstream from the Site show no difference in cyanide and fluoride concentrations which indicates that the Alcoa Site is not a significant source of these contaminants to the Columbia River. Washington Dept. of Ecology estimates that seepage of contaminated groundwater from the Alcoa Site into the Columbia River would add 0.001 ppb fluoride and 0.000008 ppb cyanide seepage — minimal levels of fluoride and cyanide — to Columbia River water.

Prior to remediation, the preliminary environmental pathways of concern related to the potliner waste piles were groundwater contamination and on-site soils. Removal of spent potliner material and insulation from the site and capping the area of contaminated soil has eliminated potential surface exposure to contaminated soil and significantly reduced the source of groundwater contamination. Four years of groundwater monitoring following the remedial action indicate that concentrations of cyanide and fluoride have exceeded MCLs in the groundwater under the contaminated soil at certain times. Groundwater samples taken where the groundwater enters the Columbia River show no detections of cyanide or fluoride. Washington Dept. of Ecology does not believe that the drinking water well one mile upgradient of the Site is threatened because the groundwater is not expected to move upstream. Monitoring data in the upgradient industrial production wells indicate that fluoride and cyanide levels are below acceptable drinking water levels or MCLs; however, some monitoring wells upgradient, within 600 feet, of the capped area have shown exceedances of MCLs for cyanide and fluoride. All pathways by which environmental receptors could potentially be exposed to Site-related contaminants have been eliminated. Since hazardous substances remain on Site, operation and maintenance activities for the cap will continue, use of the Site has been restricted, and institutional controls will remain in effect (e.g., restricted access to the Site). A long-term groundwater monitoring program has been implemented at the Site. In addition, the Site will continue to be subject to periodic five-year reviews to ensure that the remedy remains protective of human health and the environment.

The NPL site was defined as the potliner waste pile area and any contamination associated with the potliner waste (e.g., cyanide and fluoride-contaminated soil and groundwater). However, some other areas of the facility were contaminated and have been addressed, separately from the NPL site, pursuant to the Model Toxics Control Act (MTCA) or the State Dangerous Waste Law. These areas include: (1) A landfill area containing TCE contamination; (2) a lagoon area containing PCBs; (3) PCB and PAH soil contamination in the Rod Mill building; (4) PCB and TPH-contamination in a parking lot; (5) TPH and cyanide in a barge sludge lagoon; and (6) as a RCRA closure, tank sludge from the VANEXCO anodizing plant.

One of the three criteria for deletion specifies that EPA may delete a site from the NPL if ``responsible parties or other persons have implemented all appropriate response actions required''. EPA, with the concurrence of Washington Dept. of Ecology, has determined that this criteria for deletion has been met. EPA and Washington Dept. of Ecology believe that no significant threat to human health or the environment remains, because pathways of concern for exposure to contaminants no longer exist. Groundwater data show that MCLs are not exceeded at the point where groundwater from the Site enters the Columbia River and there are no drinking water wells within the area of groundwater contamination nor will any be allowed in the future. Because of the limited extent of the contaminated plume, the completed source removal, the placement of institutional controls, the technical infeasibility and lack of effectiveness of a more aggressive groundwater remedial action, and the lack of impact on the Columbia River, EPA and Washington Dept. of Ecology believe that natural attenuation over time will reduce the level of cyanide and fluoride concentrations in the groundwater under the Site. Groundwater monitoring will continue until there are no exceedances of MCLs. If new information comes available that indicates that there is a significant threat to human health or the environment then EPA or Washington Dept. of Ecology can require or conduct additional remedial action, if appropriate. Subsequently, EPA is proposing deletion of this Site from the NPL.

3. McChord Air Force Base Washrack Treatment Area, WA, Region 10.
EPA ID#WA85870024200
Date listed on the NPL: July, 1987
Date of Final Remedy Selection: 1996

McChord Air Force Base is an active 4,616-acre military installation located seven miles south of downtown Tacoma. The Washrack Treatment Area (WTA), a 22-acre area where airplanes were washed and drained of fuel, is located within the northern industrial and operational portion of the base along the western portion of the instrument runway. The site includes the former washrack (now inactive), two leach pits (now backfilled), an oil/water separator (skimmer), storm drainage infiltration ditches (now backfilled) and a layer of floating fuel on shallow groundwater in the vicinity.

The two Department of Defense Installation Restoration Program (IRP) sites that comprise the WTA (SD-54, the leach pits; and DP-60, infiltration ditches) were originally identified during the 1982 Phase I record search (CH2MHIll. 1982) conducted by McChord. The phase two IRP investigation (SAIC, 1985) measured low level organic contamination at Site DP-60 and the adjacent IRP Site SD-54. As a result of the IRP record search and investigation, further studies were recommended to confirm contaminant characteristics and distribution. The EPA designated Site SD-54 as the Washrack Treatment Area in 1984 and nominated it for inclusion on the NPL. The site was listed in 1987. In 1989 the Air Force entered into a three-party Federal Facilities Agreement (FFA) with Region 10 of the EPA and Washington

Appendix 3 — Excerpts From NPL Site RODs 211

Dept. of Ecology for conducting an investigation and cleanup of contaminants posing an unacceptable risk to human health and the environment.

A remedial investigation, which was completed in 1992, investigated source areas for the floating fuel and evaluated the nature and extent of contamination in all potentially affected media. Based on evaluation of the RI and the baseline risk assessment, the EPA determined and documented in the Record of Decision for the WTA that no remedial action under CERCLA was necessary for soil, surface water or sediment to ensure protection of human health and the environment. The ROD selected passive removal of the floating fuel to address the unacceptable risk posed by benzene associated with the floating fuel layer, and monitoring to evaluate the need for remediation of the residual fuel in the soil.

A remedial design pilot study for recovery of the floating fuel or NAPL was performed in 1993 and 1994 to determine if the layer of floating fuel could be removed. The NAPL Pilot Test Study (EA Engineering, 1994) concluded that passive removal of the fuel was not feasible due to the small amount of fuel present and that original estimates of fuel available for recovery were overestimates. The study also concluded that the soil was not a significant continuing source of contamination to groundwater and that there is an active population of bacteria present in the soil capable of naturally degrading the petroleum.

In light of the findings of the Pilot Study, an Explanation of Significant Differences (ESD) was prepared. The ESD described the results of the pilot study and the changes that were made to the ROD as a result. The ESD changed the final remedy to a combination of natural attenuation and long-term monitoring of the groundwater. Natural attenuation consists in part of allowing the hydrocarbons in the shallow groundwater to be consumed by the naturally occurring bacteria present at the site, and to allow the lighter portions of the hydrocarbons to volatilize. The shallow groundwater below the floating fuel would be monitored, as well as the shallow groundwater up-and-downgradient of the floating fuel.

The installation of one test trench and ten test pit observation wells as part of the pilot test for the passive removal of the floating fuel constituted the only active remedial action that occurred at the site. EPA concurred in a March 1995 addendum to the ROD that no further active remedial response under CERCLA is necessary at the WTA. This addendum served to signify construction completion.

Eleven rounds of groundwater samples have been collected at the floating fuel area since September 1990. All of the groundwater samples were analyzed for the six compounds for which Remedial Action Objectives (RAOs) were established in the ROD. With the exception of total petroleum hydrocarbons (TPH), levels of these compounds detected in the seven rounds conducted since completion of the ROD have been consistently below the RAOs. Semi-annual monitoring reports conducted since the ROD for the WTA are available in the site repository. One of the three criteria for deletion specifies that EPA may delete a site from the NPL if "responsible parties or other persons have implemented all appropriate response actions required". EPA, with the concurrence of Washington Dept. of Ecology, believes that this criterion for deletion has been met. Groundwater data from the Site confirm that the ROD cleanup goals have been achieved. It is concluded that there is no significant threat to human health or the environment and, therefore, no further remedial action is necessary. Subsequently, EPA is proposing deletion of this Site from the NPL.

4. Oak Grove Sanitary Landfill, MN, Region V.
EPA ID# MND980904056
Date Listed on the NPL: June, 1986
Date of Final Remedy Selection: 1996

The 45-acre Oak Grove Sanitary Landfill is a former municipal and industrial solid waste landfill in Oak Grove Township, Anoka County, Minnesota. Land consists of low regions of uplands and sand dunes interspersed among numerous lakes and wetlands. The nearby developed land use in the area is agricultural and residential. The site overlies two aquifers, which are separated by a semi-confining layer. The deeper aquifer provides regional potable water and supplies many area residential wells. Landfill operations began in 1967 and continued until 1984, when the operating license was suspended. An estimated 2.5 million cubic yards of waste is present in the landfill including acidic oil sludge, paint and solvent waste, foundry sands and sludge, inorganic acids, metal sludge, and chlorinated and unchlorinated organic compounds from pesticide manufacturing. In addition, lime sludge was used as a cover material on two thirds of the landfill.

A 1988 Record of Decision addressed the sources of contamination by containing the onsite waste and contaminated soil with a cover. EPA investigations in 1989 determined that the contaminated shallow aquifer discharges directly to the surface water of the adjoining wetlands where groundwater contamination is being reduced by natural attenuation, and thus, limiting migration of contaminants to the surface water. The present ROD addresses remediation of contaminated shallow groundwater, prevention of significant impacts on surface water from the discharge of contaminated shallow groundwater, and provides for continued use of the deep aquifer as a drinking water supply. The primary contaminants of concern affecting the groundwater are VOCs including benzene, toluene, and xylenes; and metals including arsenic.

On October 15, 1990, the Remedial Investigation/Feasibility Study and the Proposed Plan for the Oak Grove Sanitary Landfill Site were released to the public for comment. The selected remedial action for this site includes long term monitoring of the shallow and deep aquifers, surface water, and sediment at a frequency of three times per year for the first year and semi-annually thereafter; natural attenuation of shallow groundwater; abandoning non-essential wells; and implementing institutional controls including groundwater use restrictions.

During Phase 1 of the Remedial Action, debris was removed from the site and a security fence was installed around the perimeter of the Landfill. Warnings signs were posted along the fence to provide site information as well as telephone number for further information. This was completed by August 1993. Phase II began and consisted of soil excavation, installation of monitoring wells, groundwater, surface water, and sediment sampling; air monitoring, and construction of the Landfill Cover. The process began approximately on August 1992 and final inspection was completed on September 2, 1993, by representatives of MPCA and EPA.

In 1994, the Legislature of the State of Minnesota enacted the Landfill Cleanup Law, Minnesota Laws 1994, ch. 639, codified at Minnesota Stat. Sec. Sec. 115B.39 to 115B.46, authorizing the Commissioner of the Minnesota Pollution Control Agency (MPCA) to assume responsibility for future environmental response actions at qualified landfills that have receive notices of compliance from the Commissioner of MPCA. Additionally, the Act established funds to enable the MPCA to perform all necessary response, operation and maintenance at such landfills. At sites where no response for issuing a notice of compliance, all work would be expected, (under a state order or under state closure requirements) to be completed.

A notice of compliance was issued by MPCA for the Oak Grove Sanitary Landfill on May 14, 1996. MPCA has since assumed all responsibility for the Oak Grove Sanitary Landfill under the Act. Therefore, no further response actions under CERCLA are appropriate at this time. Consequently, U.S. EPA proposes to delete the site from the NPL. EPA, with concurrence of the State of Minnesota has determined that all appropriate Fund-financed responses under CERCLA at the Oak Grove Sanitary Landfill Site have been completed, and no further Superfund response is appropriate in order to provide protection of human health and the environment. Therefore, it is proposed that the site be deleted from the NPL.

5. Agrico Chemical Site, FL, Region 4.
EPA ID# FLD980221857
Date Listed on the NPL: October, 1989
Date of Final Remedy Selection: 1996

The 30-acre Agrico Chemical Co. site is bordered on the north by undeveloped land that is used for recreational purposes, on the east by interstate 110, on the south by Fairfield Drive, and on the west by CSX Transportation tracks. Industrial activity on the site began in 1889, when a company started producing sulfuric acid from iron pyrite. Around 1920, the American Agriculture Chemical Company began making fertilizer from phosphate rock. The plant underwent numerous ownership changes and its name was changed to Agrico. In 1975, Agrico stopped production, tore down the buildings, and sold the land. All that remains on the site are the foundations of five buildings, including a plant where phosphate was processed. Four ponds that were used to store liquid manufacturing wastes lie to the north and east of the ruins. In 1958, a municipal water well 1¼ miles from the site was closed due to high acidity and fluoride concentrations. The primary aquifer under the site is highly permeable, which facilitates the movement of contaminants into the groundwater. Given the direction of the flow, any contamination could enter Bayou Texar or Pensacola Bay. Thirteen county wells serving approximately 114,000 people lie within 3 miles of the site. Few residents live in the immediate vicinity of the site.

The groundwater, soil, and surface water are contaminated with fluoride, lead and arsenic. Investigations have revealed that the groundwater plume associated with the site is impacting Bayou Texar, a bayou located approximately 1 mile downgradient from the site. Residences in the immediate vicinity of the site are connected to the city water supply and are at little risk of contamination.

This site is being addressed in two long-term remedial phases focusing on source control and water pollution. An investigation of on-site soils was completed in 1992. The selected remedy includes: solidification/stabilization of soils; construction of a multi-media cap over the affected areas; and the construction of a slurry wall. Following completion of technical designs of the remedy, construction began in late 1994, and is expected to be completed in late 1996.

Under EPA oversight, the parties potentially responsible for site contamination completed an intensive study of site problems. The selected remedy to address the groundwater contamination includes: groundwater monitoring, surveying of irrigation wells, surface water monitoring of affected bayou advisory programs, establishing institutional controls, and plugging four abandoned irrigation wells with owners' permission.

After adding this site to the NPL, the EPA performed preliminary investigations and determined that no immediate actions were needed at the Agrico Chemical Co. site while cleanup actions are being planned and constructed.

6. Dover Air Force Base
Kent County, Dover, Delaware
EPA ID# DE8570024010
EPA REGION 3
Listed on NPL: March 13, 1989

The 3,700-acre Dover Air Force Base (AFB) site is the base of operation for the 436th Military Airlift Wing. The base contains 23 areas on site that were used to dispose of industrial waste. An estimated 23,000 cubic feet of waste were disposed of from 1951 to 1970. The base's operations generated numerous wastes, some in drums, including paints, solvents, waste fuels, and oil. These wastes were disposed of in various on-base locations including 12 landfills and three fire training areas. All disposal sites are earth-covered to a depth of 3 feet, with the exception of the construction debris landfill. Access to the site is restricted. There are approximately 1,000 people living on base, and 39,000 people living within a 3-mile radius of the site. The distance from the base to the nearest residence is about 1 mile, and the site is located in a commercial and residential area that is densely populated. The base well system serves about 10,000 people and is routinely monitored by the Air Force. Contaminants have not been found in this system.

Shallow on-site groundwater is contaminated with heavy metals including arsenic and cadmium and volatile organic compounds (VOCs) from former waste disposal practices. A variety of VOCs have been detected in off-site groundwater including TCE, PCE, and carbon tetrachloride. VOCs also have been detected in the sediments. VOCs and heavy metals including mercury, chromium, and cadmium have been detected in on-site stream waters. Potential health threats include exposure to contaminated groundwater used for drinking water and ingestion of contaminated

Appendix 3 — Excerpts From NPL Site RODs   215

fish and wildlife. Direct contact with contaminated surface water or sediments during recreational or site activities by area residents and workers also is a concern. A nearby freshwater wetlands is threatened by site contamination.

This site is being addressed in four stages: initial actions and three long-term remedial phases focusing on cleanup of Fire Training Area #3, the Industrial Area, and on-site groundwater. *Initial Actions:* The Air Force has cleaned up the industrial waste basins and a drum site, and has provided an alternate water supply to affected residents. A landfill and some hazardous waste areas were excavated during the runway extension in 1988 and 1989.

Fire Training Area #3: Cleanup activities began early in 1992. An underground storage tank, oil/water separator, and all associated piping were removed from the site. Contaminated soils surrounding the tanks and underlying the fire training pit also were removed. The area was capped in the fall of 1993.

Industrial Area: In 1990, the Air Force began an investigation into the nature and extent of contamination and to identify cleanup alternatives. The area is comprised of source areas including treatment units, buildings, hangars, and industrial sewer lines that are close together. The Air Force has proposed cleanup plans for source control and the removal of contamination floating on the water table, as well as the removal of contaminated soils. A final decision on the remedy is expected in 1995.

On-Site Groundwater: Groundwater monitoring currently is underway as part of the ongoing site studies. In 1990, the Air Force began an investigation into the extent and nature of groundwater contamination. A decision on the remedy was made in mid-1994, which includes removal of the sources of contamination. The design activities are completed, and the cleanup activities are scheduled to be completed in 1996.

Site Facts: The EPA, the Air Force, and the State of Delaware have entered into an Interagency Agreement (IAG) for comprehensive cleanup and compliance with Federal standards. The Dover Air Force Base also is participating in the Installation Restoration Program, a specially funded program established by the Department of Defense (DOD) in 1978 to identify, investigate, and control the migration of hazardous contaminants at military and other DOD facilities.

By cleaning up the industrial waste basins and drum sites, providing an alternate water supply to residents and workers at the Base, and cleaning up the Fire Training Areas, the Air Force has reduced the risk of immediate threats from the Dover Air Force Base site while further cleanup activities are being conducted.

7. Bendix Corporation/Allied Automotive
   Berrien County, St. Joseph, Michigan
   EPA ID# MID005107222
   EPA REGION 5
   Listed on NPL: February 21, 1990

Bendix Corporation/Allied Automotive manufactures automotive brake systems at this 36-acre site. From 1966 to 1975, a seepage lagoon on site was used for the disposal of machine-shop process wastewater. Chlorinated organic solvents, wastewater from electroplating operations, plating bath solutions, chromium, and lead may have been placed in the seepage lagoon. The lagoon was closed and capped in 1978. A private well located 750 feet from the site was closed in 1982 because of contamination. Approximately 4,300 people obtain drinking water from private wells located within 3 miles of the site.

Groundwater is contaminated with various volatile organic compounds (VOCs). People may be at risk if they drink or come into direct contact with contaminated groundwater; however, groundwater is not currently used as a source of drinking water. This site is being addressed in a long-term remedial phase focusing on groundwater cleanup.

Groundwater: The potentially responsible party, Bendix Corporation/Allied Automotive, is conducting an investigation into the nature and extent of groundwater contamination. The investigation is being conducted in two phases. Phase 1 consists of groundwater, surface water, and sediment sampling and analysis. Phase 2 consists of a soil gas survey, soil sampling, and additional groundwater sampling and analysis to locate the source of contamination. The investigation has located two apparent hot spot areas. One of these hot spots consists of shallow soil contamination. At the conclusion of the investigation, scheduled for mid-1996, a final cleanup remedy was to be selected.

Site Facts: In 1989, the EPA entered into an Administrative Order with Bendix Corporation/Allied Automotive requiring the company to conduct an investigation of site contamination. Allied Signal will seek to demonstrate a proprietary bioremediation technology in the area under the auspices of the EPA's SITE (Superfund Innovative Technology Evaluation) program. After adding this site to the NPL, the EPA performed preliminary investigations and determined that no immediate actions were required at the Bendix Corporation/Allied Automotive site while investigations continue and final cleanup activities are being planned.

Appendix 4 — WWW Sources (Web resources we found useful)

EPA

EPA's downloadable bioremediation documents
http://www.epa.gov/ORD/WebPubs/biorem

Downloadable EPA bioremediation documents — includes the Proceedings of the 1994 Natural Attenuation Meeting in Denver
http://www.epa.gov/ORD/WebPubs/biorem/

EPA's Superfund homepage
http://www.epa.gov/superfund/

EPA software and environmental databases
http://www.epa.gov/epahome/Data.html

Exposure assessment modeling information
ftp://ftp.epa.gov/epa_ceam/wwwhtml/ceamhome.htm

EPA report on natural attenuation of hexavalent chromium
http://clu-in.com/gwchexa.ht

Miscellaneous EPA datasets
http://epaserver.ciesin.org/national/epaorg/epadata.html

EPA homepage giving information on innovative cleanup technologies
http://www.clu-in.com/

DOE

DOE contamination references
http://www.er.doe.gov/production/grants/fr97_03.html

Summary of DOE contaminants
www.er.doe.gov/production/oher/EPR/pub_epr.html

DOE's Office of Environmental Management Homepage — gives ideal of remediation technologies being explored by DOE
http://www.em.doe.gov/

SEARCHABLE MAILING LISTS

Searchable archives for groundwater mailing list
http://www.reference.com/cgi-bin/pn/listarch?list = groundwater@ias.champlain.edu

Searches of the bioremediation mailing list
http://gwrp.cciw.ca/internet/bioremediation/biorem-archive.html

DOWNLOADABLE SOFTWARE

Downloadable fate and transport models
http://h2o.usgs.gov/software/

MINTEQ info and user group
ftp://aqua.ccwr.ac.za/pub/minteqa2/home.html

Downloadable version of the water chemistry program mineql
http://www.agate.net/~ersoftwr/mineql.html

Waterloo Hydrologic site contains environmental, primarily hydrologic, software,
http://www.flowpath.com

REPORTS

The Remediation Technology Development Forum. The RTDF protocol for natural attenuation of chlorinated organics is in the public area, in the bioremediation section.
http://www.icubed.com/rtdf/html/

USGS reports on MTBE
http://wwwsd.cr.usgs.gov/nawqa/pubs/ofr/ofr95.456/ofr.html
http://wwwsd.cr.usgs.gov/nawqa/pubs/wrir/wrir96.4145/wrir.doc.html

MTBE health effects report
http://www1.whitehouse.gov/WH/EOP/OSTP/NSTC/html/MTBE/mtbe-top.html

The full description of the TCLP can be found by searching the code of federal regulations on-line.
http://www.gnet.org/gnet/gov/law/cfr.htm

The downloadable LLNL report on LUFTs
http://www.llnl.gov/environment/erd/rice/

and a rebuttal
http://www.monitor.net/monitor/free/luftcleanup.html

Appendix 4 — WWW Sources

MISCELLANEOUS

The American Society for Testing and Materials home page
http://www.astm.org/

On-line Code of Federal Regulations
http://www.gnet.org/gnet/gov/law/cfr.htm

A continually updated monitor of remediation technologies (legal/reg/science)
http://www.gvi.net/soils/index2.html

A listing of bioremediation resources
http://gwrp.cciw.ca/internet/bioremediation/

European environmental information
http://www.ovam.be/internetrefs/english.htm
http://www.ulb.ac.be/ceese/cds.html

An extensive collection of ecological Models
http://dino.wiz.uni-kassel.de/model_db/server.html

On-line chemical compound information
http://chemfinder.camsoft.com/

The Superfund factbook
http://www.cnie.org/nle/waste-1.html

The Texas Natural Attenuation Protocol
http://www.tnrcc.state.tx.us/waste/pst/index.html

References

Adamson A. W. (1990) *Physical Chemistry of Surfaces*. Wiley-Interscience, New York.

Adriaens P., Fu Q., and Grbic-Galic D. (1995) Bioavailibility and transformation of highly chlorinated dibenzo-p-dioxins and dibenzofurans in anaerobic soils and sediments. *Environ. Sci. Technol.* **29**, 252–2260.

AFCEE. (1992) Environmental chemistry function Installation Restoration Program analytical protocols. U.S. Air Force Center for Environmental Excellence.

AFCEE. (1993) *Handbook to support installation restoration program (IRP) remedial investigations and feasibility studies (RI/FS)*. U.S. Air Force Center for Environmental Excellence.

Aggarwal P. K. and Hinchee R. E. (1991) Monitoring *in situ* biodegradation of hydrocarbons using stable carbon isotopes. *Environ. Sci. Technol.* **25**, 1173–1180.

Alexander J. H. (1993) *In Defense of Garbage*. Praeger Press, Westport, CT.

Alexander M. (1995) How toxic are toxic chemicals in soil? *Environ. Sci. Technol.* **29**, 2713–2717.

Alpers C. N. and Blowes D. W. (1994) *Environmental Geochemistry of Sulfide Oxidation*. American Chemical Society.

Alshawabkeh A. N. and Acar Y. B. (1992) Removal of contaminants from soils by electrokinetics: A theoretical treatise. *J. Environ. Sci. Health* **A27**, 1835–1861.

American Public Health Association. (1992) *Standard Methods for the Examination of Water and Wastewater, 18th ed.*

American Society for Testing and Materials. (1995) Standard guide for risk-based corrective action at petroleum-release sites. ASTM.

Ames B. N. and Gold L. S. (1991) Cancer prevention strategies greatly exaggerate risks. *Chem. Eng. News* **Jan. 7**.

Baedecker M. J., Cozzarelli I. M., Eganhouse R. P., Siegel D. I., and Bennett P. C. (1993) Crude oil in a shallow sand and gravel aquifer. 3. Biogeochemical reactions and mass balance modeling in anoxic groundwater. *App. Geochem.* **8**, 569–586.

Baedecker M. J., Siegel D. I., Bennett P. C., and Cozzarelli I. M. (1988) The fate and effects of crude oil in a shallow aquifer: I. The distribution of chemical species and geochemical facies. *U.S. Geological Survey Toxic Substances Hydrology Program*.

Bagchi A. (1987) Natural attenuation mechanisms of landfill leachate and effects of various factors on the mechanisms. *Waste Management and Research* **5**, 453–464.

Balsley S. D., Brady P. V., Krumhansl J. L., and Anderson H. L. (1996) Iodide retention by metal sulfide surfaces: cinnabar (HgS) and chalcocite (Cu_2S). *Environ. Sci. Technol.* **30**, 3025–3027.

Barcelona M. J. and Holm T. R. (1991) Oxidation-reduction capacities of aquifer solids. *Environ. Sci. Technol.* **25**, 1565–1572.

Barker J. F., Patrick G. C., and Major D. (1987) Natural attenuation of aromatic hydrocarbons in a shallow sand aquifer. *Groundwater Monitoring and Review* **winter**, 64–71.

Barker J. F., Tessmann J. S., Plotz P. E., and Reinhard M. (1986) The organic geochemistry of a sanitary landfill leachate plume. *J. Contam. Hydrol.* **1**, 171–189.

Barney G. S. (1984) Radionuclide sorption and desorption reactions with interbed materials from the Columbia River Basalt formation. In *Geochemical Behavior of Radioactive Waste*, pp. 3–23. American Chemical Society.

Bartlett R. J. and James B. R. (1988) Mobility and bioavailability of chromium in soils. In *Chromium in the Natural and Human Environments*, Vol. 20 (J. O. Nriagu and E. Nieboer, Eds.), pp. 267–307, John Wiley & Sons, New York.

Bear J., Nichols E., Ziago J., and Kulsretha A. (1994) Effect of Contaminant Diffusion into and out of Low-Permeability Zones. *Lawrence Livermore National Labratory (UCRL-ID-115625)*.

Begley R. (1996) Risk-based remediation guidelines take hold. *Environ. Sci. Technol.* **30**, 438a–441a.

Bennett P. C., Siegel D. I., Baedecker M. J., Cozzarelli I. M., and Hult M. F. (1993) Crude oil in a shallow sand and gravel aquifer, I. Hydrogeology and inorganic geochemistry. *Appl. Geochm.* **8**, 529–549.

Berner R. A., Sjoberg E. L., Velbel M. A., and Krom M. D. (1980) Dissolution of pyroxenes and amphiboles during weathering. *Science* **207**, 1205–1207.

Berner U. (1992) Evolution of pore water chemistry during degradation of cement in a radioactive waste repository environment. *Waste Management* **12**, 201–219.

Bethke C. M. (1994) *The Geochemist's Workbench™. A Users Guide to Rxn, Act2, Tact, React, and Gtplot.* University of Illinois.

Bjerg P. L., Rugge K., Pedersen J. K., and Christensen T. H. (1995) Distribution of redox-sensitive groundwater quality parameters downgradient of a landfill (Grindsted, Denmark). *Environ, Sci. Technol.* **29**, 1387–1394.

Bobertz B. C. (1996) Transferring the blame. *The Environmental Forum* **January/February**, 22–31.

Borden R. C., Gomez C. A., and Becker M. T. (1994) Geochemical indicators of intrinsic bioremediation. *Ground Water* **33**, 180–189.

Borgert C. J., Roberts S. M., Harbison R. D., and James R. C. (1995) Influence of soil half-life on risk assesment of carcinogens. *Reg. Toxic. and Pharmacol* **22**, 143–151.

Bossert I. and Bartha R. (1984) The fate of petroleum in soil ecosystems. In *Petroleum Microbiology* (R. M. Atlas, Ed.). Macmillan, New York.

Boulding J. R. (1995) *Practical Handbook of Soil, Vadose Zone, and Groundwater Contamination.* CRC Press, Boca Raton, FL.

Bradley P. M. and Chapelle F. H. (1995) Factors affecting microbial 2,4,6-Trinitrotoluene mineralization in contaminated soil. *Environ. Sci. Technol.* **29**, 802–806.

Bradley P. M. and Chapelle F. H. (1996) Anaerobic mineralization of vinyl-chloride in Fe(III)-reducing, aquifer sediments. *Environ. Sci. Technol.* **30**, 2084–2086.

Bradley P. M. and Chappelle F. H. (1995) Rapid toluene mineralization by aquifer microorganisms at Adak, Alaska: Implication for intrinsic bioremediation in cold environments. *Environ. Sci. Technol* **29**, 2778–2781.

Brady P. V. (1996) Natural attenuation of metals by irreversible sorption. *IBC's Second Annual International Symposium on Intrinsic Bioremediation.*

Brady P. V. and House W. A. (1996) Surface-controlled dissolution and growth of minerals. In *Physics and Chemistry of Mineral Surfaces* (P. V. Brady, Ed.), pp. 225–306. CRC Press, Boca Raton, FL.

Brady P. V. and Kozak M. W. (1995) Geochemical engineering of low level radioactive waste in cementitous environments. *Waste Management* **15**, 293–301.

Brady P. V. and Walther J. V. (1989) Controls on silicate dissolution rates in neutral and basic pH solutions at 25°C. *Geochim. Cosmochim. Acta* **53**, 2823–2830.

Brady P. V. and Zachara J. M. (1996) Geochemical applications of mineral surface science. In *Physics and Chemistry of Mineral Surfaces* (P. V. Brady, Ed.), pp. 307–356. CRC Press, Boca Raton, FL.

References

Brannon J. M., Pennington J. C., McFarland V. A., and Hayes C. (1995) The Effects of sediment contact time on Koc of nonpolar organic contaminants. *Chemosphere* **31**(3465–3473).

Brookins D. G. (1978) Eh-pH diagrams for elements from Z = 40 to Z = 52: applications to the OKLO natural reactor, Gabon. *Chem. Geol.* **23**, 325–342.

Brookins D. G. (1984) *Geochemical Aspects of Radioactive Waste Disposal.* Springer-Verlag, New York.

Browman M. G. and Spalding B. P. (1984) Reduction of radiostrontium mobility in acid soils by carbonate treatment. *J. Environ. Qual.* **13**, 166–172.

Brown R. A., Hicks P. M., Hicks R. J., and Leahy M. C. (1995) Postremediation bioremediation. In *Intrinsic Bioremediation* (R. E. Hinchee, J. T. Wilson, and D. C. Downey, Eds.), pp. 77–84. Battelle Press, Columbus, OH.

Bryan C., Siegel M. D., and Marozas D. C. (1996) Hydrogeologic and Geochemical Data Calculation and Analysis Notebook — Riverton, Wyoming. Sandia National Laboratories.

Buchanan R. J. J. (1996) Intrinsic bioremediation of chlorinated organics. *IBC's 2nd International Symposium on Bioremediation: Natural Attenuation.*

Bunzl K., Flessa H., Kracke W., and Schimmack W. (1995) Association of fallout $^{239+240}$Pu and ^{241}Am with various soil components in successive layers of a grassland soil. *Environ. Sci. Technol.* **29**, 2513–2518.

Bushcheck T. E. and Alcantar C. M. (1995) Regression techniques and analytical solutions to demonstrate intrinsic bioremediation. In *Intrinsic Bioremediation* (R. E. Hinchee, J. T. Wilson, and D. C. Downey, Eds.), pp. 109–116. Battelle Press, Columbus, OH.

Bushcheck T. E., O'Reilly K. T., and Nelson S. N. (1993) Evaluation of intrinsic bioremediation at field sites. *The 1993 Symposium on Petroleum Hydrocarbons and Organic Chemicals in Groundwater: Prevention, Detection, and Restoration,* 367–381.

Caldwell K. R., Tarbox D. L., Barr D., Fiorenza S., Dunlap L. E., and Thomas S. B. (1992) Assessment of natural bioremediation as an alternative to traditional active remediation at selected Amoco Oil Company sites, Florida. *The 1992 Symposium on Petroleum Hydrocarbons and Organic Chemicals in Ground Water: Prevention, Detection, and Restoration,* 509–525.

Chaineau C. H., Morel J. L., and Oudout J. (1995) Microbial degradation in soil microcosms of fuel oil hydrocarbons from drilling cuttings. *Environ. Sci. Technol.* **29**, 1615–1621.

Chapelle F. H. (1993) *Groundwater Microbiology and Geochemistry.* John Wiley & Sons, New York.

Chapelle F. H. (1996) Identifying redox conditions that favor the natural attenuation of chlorinated ethenes in contaminated groundwater systems. *Second Annual International Symposium on Intrinsic Bioremediation: Natural Attenuation,* Annapolis, MD, Dec. 3–4, 1996. International Business Communications, Southborough, MD.

Chiang C. Y., Salanitro J. P., Chai E. Y., Colhart J. D., and Klein C. L. (1989) Aerobic biodegradation of benzene, toluene, and xylene in a sandy aquifer — data analysis and computer modeling. *Ground Water* **27**, 823–834.

Chou L., Garrels R. M., and Wollast R. (1989) Comparative study of the kinetics and mechanisms of dissolution of carbonate minerals. *Chem. Geol.* **78**, 269–282.

Cline P. V. and Viste D. (1985) Migration and degradation patterns of volatile organic compounds. *Waste Management & Research* **3**, 351–360.

Cochran M. F. and Berner R. A. (1993) Enhancement of silicate weathering rates by vascular land plants: quantifying the effect. *Chemical Geology* **107**, 213–215.

Cohen B. A., Krumholz L. R., Kim H., and Hemond H. F. (1995) In-situ biodegradation of toluene in a contaminated stream. 2. laboratory study. *Environ. Sci. Technol.* **29**, 117–125.

Cole G. M. (1994) *Assessment and Remediation of Petroleum Contaminated Sites.* Lewis Publishers/CRC Press, Boca Raton, FL.

Coleman W. E., Munch J. W., Streicher R. P., Ringhand H. P., and Kopfler F. C. (1984) *Environ. Contam. Toxicol.* **13**, 171–178.

Comans R. N. J., Haller M., and DePreter P. (1991) Sorption of cesium on illite: Nonequilibrium behavior and reversibility. *Geochim. Cosmochim. Acta.* **55**, 433–440.

Cooney C. M. (1996) EPA nears completion of "natural attenuation" remediation policy. *Environ. Sci. Technol.* **30**, 478A.

Costa V., Boopathy R., and Manning J. (1996) Isolation and characterization of a sulfate-reducing bacterium that removed TNT (2,4,6-Trinitrotoluene) under sulfate reducing and nitrate-reducing conditions. *Biores. Technol.* **56**, 273–278.

Cox E., Edwards E., Lehmicke, L. Major, D. (1995) Intrinsic biodegradation of trichloroethene from geochemical data. In *Intrinsic Bioremediation* (R. E. Hinchee, J. T. Wilson, and D. C. Downey, Eds.). Battelle Press, Columbus, OH.

Cross F. L. J. and Robinson R. (1989) Infectious waste. In *Standard Handbook of Hazardous Waste Treatment and Disposal* (H. M. Freeman, Ed.), pp. 4.35–4.45. McGraw-Hill, New York.

Davis A., Ruby M. V., Bloom M., Schoof R., Freeman G., and Bergstrom P. D. (1996) Mineralogic constraints on the bioavailability of arsenic in smelter-impacted soils. *Environ. Sci. Technol.*, 392–399.

Davis J. A., Fuller C. C., and Cook A. D. (1987) A model for trace metal sorption processes at the calcite surface: Adsorption of Cd^{2+} and subsequent solid solution formation. *Geochim. Cosmochim. Acta.* **51**, 1477–1490.

Davis J. A. and Kent D. B. (1990) Surface complexation modeling in aqueous chemistry. In *Mineral-Interface Geochemistry* (M. F. Hochella and A. F. White, Eds.), pp. 177–260, Mineralogical Society of America, Washington, D.C.

Davis J. W. and Carpenter C. L. (1990) Aerobic biodegradation of vinyl chloride in groundwater samples. *Appl. Environ. Microbiol.* **56**, 3878–3880.

Davis J. W., Klier N. J., and Carpenter C. L. (1994) Natural biological attenuation of benzene in groundwater beneath a manufacturing facility. *Ground Water* **32**, 215–226.

Delzer G. C., Zogorski J. S., Lopes T. J., and Bosshart R. L. (1996) Occurrence of the Gasoline Oxygenate MTBE and BTEX compounds in urban stormwater in the United States, 1991–1995. USGS.

Deng B. and Stone A. T. (1996) Surface-catalyzed chromium(VI) reduction: reactivity comparisons of different organic reductants and different oxide surfaces. *Environ. Sci. Technol.* **30**, 2484–2494.

Dolfing J., Zeyer J., Binder-Eicher P., and Schwarzenbach R. P. (1990) Isolation and characterization of a bacterium that mineralizes toluene in the absence of molecular oxygen. *Arch. Microbiol.* **154**, 336–341.

Domenico P. A. (1987) An analytical model for multidimensional transport of a decaying contaminant species. *J. Hydrology* **91**, 49–58.

Domenico P. A. and Schwartz F. (1990) *Physical and Chemical Hydrogeology.* Wiley-Interscience, New York.

Dorn R. I. and Brady P. V. (1995) Rock-based measurement of temperature-dependent plagioclase weathering. *Geochim. Cosmochim. Acta* **59**, 2847–2852.

Doty C. B. and Travis C. C. (1991) The effectiveness of groundwater pumping as a restoration technology. University of Tennessee, Waste Management Research and Education Institute.

Dove P. M. (1994) The dissolution kinetics of quartz in sodium chloride solutions to 25°C to 300°C. *Am. J. Sci.* **294**, 665–712.

References

Dragun J. (1988) *The Soil Chemistry of Hazardous Materials.* Hazardous Materials Control Research Institute, Silver Spring, MD.

Dragun J. (1991) Geochemistry and soil chemistry reactions occurring during *in situ* vitrification. *J. Haz. Mat.* **26**, 343–364.

Dupont R. R., Sorensen D. L., and Kemblowski M. (1994) Evaluation of intrinsic bioremediation at an underground storage site. *EPA Symposium on Intrinsic Bioremediation of Ground Water*, 176–177.

Duque E., Haidour A., Godoy F., and Ramos J. L. (1993) Construction of a Pseudomonas hybrid strain that mineralizes 2,4,6-trinitrotoluene. *J. Bacteriol.* **175**, 2278–2283.

Durant N. D., Jonkers C. A. A., Wilson L. P., and Bouwer E. J. (1995a) Enhanced biodegradation of napthalene in MGC aquifer microcosms. In *Intrinsic Bioremediation* (R. E. Hinchee, J. T. Wilson, and D. C. Downey, Eds.), pp. 189–203. Battelle Press, Columbus, OH.

Durant N. D., Wilson L. P., and Bouwer E. J. (1995b) Microcosm studies of subsurface PAH-degrading bacteria from a former manufactured gas plant. *J. Contamin. Hydrol.* **17**, 213–237.

Dzombak D. A. and Morel F. M. M. (1990) *Surface Complexation Modeling: Hydrous Ferric Oxide.* John Wiley & Sons, New York.

Eary L. E. and Rai D. (1987) Kinetics of chromium(III) oxidation to chromium(VI) by reaction with manganese dioxides. *Environ. Sci. Technol.* **21**, 1187–1193.

Eckermann K. F., Wolbarst A. B., and Richardson A. C. (1988) Federal Guidance Report. In *Limiting values of radionuclide intake and air concentration and dose conversion factors for inhalation, submersion and ingestion.*

Edwards E. A. and Grbic-Galic D. (1994) Anaerobic degradation of toluene and o-xylene by a methanogenic consortium. *Appl. Environ. Microbiol.* **60**, 313–322.

Eganhouse R. P., Baedecker M. J., and Cozzarelli I. M. (1994) Biogeochemical processes in an aquifer contaminated by crude oil: An overview of studies at the Bemidji, Minnesota, Research Site. *US-EPA Symposium on Natural Attenuation of Groundwater.*

Egli T. and Bally M. (1996) How is the microbial degradation of trace compounds regulated? *EAWAG News* **40**, 23–27.

Erlich G. G., Goerlitz D. F., Godsy E. M., and Hult M. F. (1982) Degradation of phenolic contaminants in groundwater by anaerobic bacteria: St Louis Park, Minnesota. *Ground Water* **20**, 703–710.

Evans P. J., Mang D. T., and Young L. Y. (1991) Degradation of toluene and m-xylene and transformation of o-xylene by denitrifying enrichment cultures. *Appl. Environ. Microbiol.* **57**, 450–454.

Facer G. (1980) Quantities of transuranic elements in the environment from operations relating to nuclear weapons. In *Transuranic Elements in the Environment* (W. C. Hanson, Ed.), pp. 86–91. Technical Information Center/U.S. Dept. of Energy.

Faure G. (1991) *Principles and Applications of Inorganic Geochemistry.* MacMillan, New York.

Fetter C. W. (1989) Transport and fate of organic compounds in groundwater. In *Recent Advances in Groundwater Hydrology* (J. E. Moore, A. A. Zaporozec, S. C. Csallany, and T. C. Varney, Eds.), pp. 174–184. American Institute of Hydrology, Arlington, VA.

Fetter C. W. (1992) *Contaminant Hydrogeology.* Prentice-Hall, Englewood Cliffs, NJ.

Fiorenza S., Hockman E. L., Szojka S., Woeller R. M., and Wigger J. W. (1994) Natural anaerobic degradation of chlorinated solvents at a Canadian manufacturing plant. In *Bioremediation of Chlorinated and Polycyclic Aromatic Hydrocarbon Compounds* (R. E. Hinchee, A. Leeson, L. Semprini, and S. K. Ong, Eds.). Lewis Publishers/CRC Press, Boca Raton, FL.

Fish K. M. (1996) Influence of Aroclor-1242 concentration on polychlorinated biphenyl biotransformations in Hudson River test-tube microcosms. *Appl. Environ. Microbiol.*, 3014–3016.

Francis A. J. (1994) Microbial transformations of radioactive wastes and environmental restoration through bioremediation. *J. Alloys and Colloids* **213**, 226–231.

Freeze R. A. and Cherry J. A. (1979) *Groundwater*. Prentice-Hall, Englewood Cliffs, New Jersey.

Freeze R. A. and Cherry J. A. (1989) Guest editorial: What has gone wrong? *Ground Water* **27**, 458–464.

Freeze R. A., Massmann J., Smith L., Sperling T., and James B. (1990) Hydrogeological decision analysis: 1. A framework. *Ground Water* **28**, 738–766.

Fries M. R., Zhou J., Chee-Sanford J., and Tiedje J. M. (1994) Isolation, characterization, and distribution of denitrifying toluene degraders from a variety of habitat. *Appl. Environ. Microbiol.* **60**, 2802–2810.

Fuex A. N. (1977) The Use of stable carbon isotopes in hydrocarbon exploration. *J. Geochem. Explor.* **6**, 139–162.

Fuller C. C. and Davis J. A. (1987) Processes and kinetics of Cd^{2+} sorption by a calcareous aquifer sand. *Geochim. Cosmochim. Acta* **51**, 1491–1502.

Funk S. B., Crawford D. L., Crawford R. L., Mead G., and Davoshoover W. (1995) Full-scale anaerobic bioremediation of trinitrotoluene (TNT) contaminated soil: A U.S. EPA SITE program demonstration. *Appl. Biochem. Biotechnol.* **51-2**, 625–633.

GAO. (1996) Federal Facilities: Consistent Relative Risk Evaluations Needed for Prioritizing Cleanups. US-GAO.

Gershey E. L., Klein R. C., Party E., and Wilkerson A. (1990) *Low Level Radioactive Waste — From Cradle to Grave.* Van Nostrand Reinhold, New York.

Gilham R. W. and Burris D. R. (1992) *In situ* treatment walls — chemical dehalogenation, denitrification, and bioaugmentation. *Subsurface Restoration Conference.*

Godsy E. M., Goerlitz D. F., and Grbic-Galic D. (1992) Methanogenic biodegradation of creosote contaminants in natural and simulated groundwater ecosystems. *Ground Water* **30**, 232–242.

Graustein W. C., Cromack K., and Sollins P. (1977) Calcium oxalate: occurrence in soils and effect on nutrient and geochemical cycles. *Science* **198**, 1252–1254.

Gray P. (1990) Radioactive materials could pose problems for the gas industry. *Oil and Gas Journal* **June 25, 1990**.

Grbic-Galic D. (1990) Anaerobic microbial transformation of nonoxygenated aromatic and alicyclic compounds in soil, subsurface, and freshwater sediments. In *Soil Biochemistry* (J. M. Bollag and G. Stotzky, Eds.), pp. 117–189. Marcel-Dekker, New York.

Grbic-Galic D. and Vogel T. M. (1987) Transformation of toluene and benzene by mixed methanogenic cultures. *Appl. Environ. Microbiol.* **53**, 254–260.

Greenberg M. and Anderson R. (1984) Hazardous waste sites: the credibility gap. *New Brunswick N. J. Center for Urban Policy Research.*

Grütter A., Von Gunten H. R., Rossler E., and Keil R. (1994) Sorption of nickel and cobalt on a size fraction of unconsolidated glaciofluvial deposits and on clay minerals. *Radiochim. Acta* **65**, 181–187.

Guest P. R., Benson L. A., and Rainsberger T. J. (1995) Inferring biodegradation processes for trichloroethane from geochemical data. In *Intrinsic Bioremediation* (R. E. Hinchee, J. T. Wilson, and D. C. Downey, Eds.). Battelle Press, Columbus, OH.

Hadley P. W. and Armstong R. (1991) Where's the benzene — examining California groundwater quality surveys. *Groundwater* **29**, 35–40.

References

Hale J. R., Foster D. R., and Misquitta N. J. (1996) Naturally Occurring biodegradation of TCE under aerobic conditions. *IBC's Second Annual International Symposium on Intrinsic Bioremediation.*

Hamaker J. W. (1979) *Interpretation of Soil Leaching Experiments, Dynamics of Pesticides in the Environment.* Plenum Press, New York.

Hedin R. S., Watzlaf G. R., and Nairn R. W. (1994) Passive treatment of acid mine drainage with limestone. *J. Env. Quality* **23**, 1338–1345.

Heijman C. G., Greider E., Hollinger C., and Schwarzenbach R. (1995) Reduction of nitroaromatic compounds coupled to microbial iron reduction in labratory aquifer columns. *Environ. Sci. Technol.* **29**, 775–783.

Hem J. D. (1976b) Geochemical controls on lead concentrations in stream water and sediments. *Geochim. Cosmochim. Acta.* **40**, 599–609.

Hem J. D. (1989) Study and interpretation of the chemical characteristics of natural water. *United States Geol. Surv. Water Supply Paper 2254.*

Henderson T. (1994) Geochemical reduction of hexavalent chromium in the Trinity Sand Aquifer. *Ground Water* **32**, 477–486.

Heron G., Bjerg P. L., and Christensen T. H. (1995) Redox buffering in shallow aquifers contaminated by leachate. In *Intrinsic Bioremediation* (R. E. Hinchee, J. T. Wilson, and D. C. Downey, Eds.), pp. 143–152. Battelle Press, Columbus, OH.

Heron J. and Christensen T. H. (1995) Impact of sediment-bound iron on redox buffering in a landfill leachate polluted aquifer (Vejen, Denmark). *Environ. Sci. Technol.* **29**, 187–192.

Holt B. D. and Sturchio N. C. (1996) High temperature method for concersion of chlorinated organics to CH_3Cl and CO_2 for isotopic analyses of chlorine and carbon. *Abs. Am. Chem. Soc.* **212(1)**, 154.

Hopkins G. D., Semprini L., and McCarty P. L. (1993) Microcosm and *in situ* field studies of enhanced biotransformation of trichloroethylene by phenol-utilizing microorganisms. *Appl. Environ. Microbiol.* **59**, 2277–2285.

Howard P. H., Boethling R. S., Jarvis W. F., Meylan W. M., and Michalenko E. M. (1991) *Handbook of Environmental Degradation Rates.* Lewis Publishers/CRC Press, Boca Raton, FL.

Howard P. H., Hueber A. E., and Boethling R. S. (1987) Biodegradation data evaluation for structure/biodegradability relations. *Environ. Toxicol. Chem.* **6**, 1–10.

Hoye R. L. and Hubbard S. J. (1989) Mining wastes. In *Standard Handbook of Hazardous Waste Treatment and Disposal* (H. M. Freeman, Ed.), pp. 4.47–4.51. McGraw-Hill, New York.

Hsu C. N. and Chang K. P. (1994) Sorption and desorption behavior of cesium on soil components. *Appl. Rad. Isotopes* **45**, 433–437.

Huang X. and Evangelou V. P. (1994) Suppression of pyrite oxidation rate by phosphate addition. In *Environmental Geochemistry of Sulfide Oxidation*, Vol. 550 (C. N. Alpers and D. W. Blowes, Eds.), pp. 562–573. American Chemical Society.

Huesemann M. H. (1995) Predictive model for estimating the extent of petroleum hydrocarbon degradation in contaminated soils. *Environ. Sci. Technol.* **29**, 7–18.

Hughes J. B., Shanks J., Vanderford M., Lauritzen J., and Bhadra R. (1997) Transformation of TNT by aquatic plants and plant tissue cultures. *Environ. Sci. Technol.* **31**, 266–271.

Hume L. A. and Rimstidt J. D. (1992) The biodurability of chrysotile asbestos. *Am. Mineral.* **77**, 1125–1128.

Huyakorn P. and Pinder G. F. (1983) *Computational Models in Subsurface Flow.* Academic Press, New York.

Isherwood W. F., Rice D. J., Ziagos J., and Nichols E. (1993) "Smart" pump-and-treat. *J. Haz. Mat.* **35**(413–426).

Jackson M. L. (1969) *Soil Chemical Analysis — Advanced Course.*

Jackson R. E., Priddle M. W., and Lesage S. (1990) Transport and fate of CFC-113 in groundwater. In *Proceedings of Petroleum Hydrocarbons and Organic Chemicals in Ground Water: Prevention, Detection, and Restoration*, pp. 129–142. National Water Well Association.

James B. R. and Bartlett R. J. (1983) Behavior of chromium in soils: VII. Adsorption and reduction of hexavalent forms. *J. Env. Qual.* **12**, 177–181.

Johnson J. W., Oelkers E. H., and Helgeson H. C. (1991) SUPCRT92: A software package for calculating the standard molal thermodynamic properties of minerals, gases, aqueous species, and reactions. Dept. of Geology and Geophysics, University of California at Berkeley.

Jones A. B. (1996a) RBCA aims to aid sites suffering from heavy metal risks. *Environmental Technology* **6**, 59–64.

Jones C. C. (1996b) Natural Attenuation: A remedial option for a dissolved phase plume at an MGP tar disposal site. *IBC's Second Annual International Symposium on Intrinsic Bioremediation.*

Kampbell D. H., Wilson J. T., and Vandegrift S. A. (1989) Dissolved oxygen and methane in water by a GC headspace equilibrium technique. *Int. J. Environ. Anal. Chem.* **36**, 249–257.

Karickhoff S. W., Brown D. S., and Scott T. (1979) Sorption of hydrophobic pollutants on natural sediments. *Water Research* **13**, 241–248.

Kemblowski M. W., Salanitro J. P., Deeley G. M., and Stanley C. C. (1987) Fate and transport of residual hydrocarbon in groundwater: a case study. *Petroleum hydrocarbons and organic chemicals in groundwater: Prevention, detection, and preservation: A conference and exposition*, 207–231.

Kenaga E. E. (1980) Predicted bioconcentration factors and soil sorption coefficients of pesticides and other chemicals. *Ecotoxicology and Environ. Safety* **4**, 26–38.

Kenaga E. E. and Goring E. A. I. (1980) Relationship between water solubility, soil sorption, octanol-water partitioning, and concentration of chemicals in biota, pp. 78–115. American Society for Testing and Materials.

Khan S. A., Riazurrehman, and Khan M. A. (1994) Sorption of cesium on bentonite. *Waste Management* **14**, 629–642.

Kim H., Hemond H. F., Krumholz L. R., and Cohen B. A. (1995) In-situ biodegradation of toluene in a contaminated stream. 1. field study. *Environ. Sci. Technol.* **29**, 108–116.

King M. W. G., Barker J. F., and Hamilton K. A. (1995) Natural attenuation of coal tar organic in groundwater. In *Intrinsic Bioremediation* (R. E. Hinchee, J. T. Wilson, and D. C. Downey, Eds.), pp. 171–179. Battelle Press, Columbus, OH.

Klecka G. M., Davis J. W., Gray D. R., and Madsen S. S. (1990) Natural bioremediation of organic contaminants in groundwater: Cliffs-Dow superfund site. *Ground Water* **28**, 534–543.

Krauskopf K. B., Van Andel T. H., and Smith P. J. (1988) *Radioactive Waste Disposal and Geology.* Chapman & Hall, New York.

Laperche V., Traina S. J., Gaddam P., and Logan T. J. (1996) Chemical and mineralogical characterizations of Pb in a contaminated soil: reactions with synthetic apatite. *Environ. Sci. Technol.* **30**, 3321–3326.

Leahy J. G. and Colewell R. R. (1990) Microbial degradation of hydrocarbons in the environment. *Microbiol. Rev.* **53**, 305–315.

Lee M. D., Mazierski P. F., Buchanan R. J. J., Ellis D. E., and Sehayek L. S. (1995) Intrinsic *in situ* anaerobic biodegradation of chlorinated solvents at an industrial landfill. In *Intrinsic Bioremediation* (R. E. Hinchee, J. T. Wilson, and D. C. Downey, Eds.), pp. 205–222. Battelle Press, Columbus, OH.

References

Lerman A. (1985) *Geochemical Processes: Water and Sediment Environments.* John Wiley & Sons, New York.

Lindgren E. R., Kozak M. W., and Mattson E. D. (1994) Electrokinetic remediation of anionic contaminants from unsaturated soils. In *Emerging Technologies in Hazardous Waste Management,* Vol. 554 (D. W. Tedder and F. G. Pohland, Eds.). American Chemical Society.

Long J. C. (1996) *Chem. & Eng. News*(9/25/96), 7–8.

Lovley D. R., Baedecker M. J., Lonergan D. J., Cozzarelli I. M., Phillips E. J. P., and Siegel D. M. (1989) Oxidation of aromatic contaminants coupled to microbial iron reduction. *Nature* **339**, 297–299.

Ludvigsen L., Heron G., Albrechtsen H.-J., and Christensen T. H. (1995) Geomicrobial and geochemical redox processes in a landfill-polluted aquifer. In *Intrinsic Degradation* (R. E. Hinchee, J. T. Wilson, and D. C. Downey, Eds.). Battelle Press, Columbus, OH.

Lybarger J. and Spengler R. (1996) Quoted in *Environ. Sci. Technol.* **30**, 429A.

Lyman W. J., Reehl W. F., and Rosenblatt D. H. (1982) *Handbook of Chemical Property Estimation Methods.* McGraw-Hill, New York.

MacDonald J. A. and Kavanaugh M. C. (1994) Restoring contaminated groundwater: An achievable goal? *Environ. Sci. Technol* **28**, 362–368A.

MacIntyre W. G., Boggs M., Antworth C. P., and Stauffer T. B. (1993) Degradation kinetics of aromatic organic solutes introduced into a heterogeneous aquifer. *Water Resources* **29**, 4045–4051.

Mackay D. M. and Cherry J. A. (1989) Groundwater contamination: pump-and-treat remediation. *Environ. Sci. Technol.* **23**, 630–636.

Madsen E. L., Sinclair J. L., and Ghiorse W. C. (1991) *In situ* biodegradation: Microbiological patterns in a contaminated aquifer. *Science* **252**, 830–833.

Major D. W., Hodgins W. W., and Butler B. J. (1991) Field and laboratory evidence for *in situ* biotransformation of tetrachlorethene to ethene and ethane at a chemical transfer facility in North Toronto. In *Onsite Bioreclamation* (R. E. Hinchee and R. F. Olfenbuttel, Eds.), pp. 147–171. Butterworth-Heinemann, Newton, MA.

Martin M. and Imbrogiotta T. E. (1994) Contamination of groundwater with trichloroethylene at the Building 24 Site at Picatinny Arsenal, New Jersey. *US-EPA Symposium on Natural Attenuation of Groundwater.*

McCarthy J. F. and Zachara J. M. (1989) Subsurface transport of contaminants. *Environ. Sci. Technol.* **23**, 496–502.

McCarty P. L. (1994) An overview of anaerobic transformation of chlorinated solvents. *U.S. EPA Symposium on Natural Attenuation of Groundwater.*

McCarty P. L. and Reinhard M. (1993) Biological and chemical transformations of halogenated aliphatic compounds in aquatic and terrestrial environments. In *The Biogeochemistry of Global Change: Radiative Trace Gases* (R. S. Oremland, Ed.). Chapman & Hall, New York.

McCarty P. L., Reinhard M., and Rittmann B. E. (1981) Trace organics in groundwater. *Environ. Sci. Technol.* **15**, 40–51.

Means J. L., Crerar D. A., and Duguid J. O. (1978) Migration of radioactive wastes: radionuclide mobilization by completing agents. *Science* **200**, 1477–1481.

Miller I., Rossisk R., and Cunnane M. (1993) A new methodology for repository site suitability evaluation. *Waste Management* **13**(494–501).

Millette D., Barker J. F., Comeau Y., Butler B. J., Frind E. O., Clement B., and Samson R. (1995) Substrate interaction during aerobic biodegradation of creosote-related compounds: a factorial batch experiment. *Environ. Sci. Technol.* **29**, 1944–1952.

Milloy S. J. (1995) *Science-based Risk Assesment.* National Environmental Policy Institute.

Morel F. M. M. and Hering J. (1993) *Principles and Applications of Aquatic Chemistry.* Wiley-Interscience, New York.

Morgan P. and Watkinson R. J. (1989) Hydrocarbon degradation in soils and methods for soil biotreatment. *CRC Critical Rev. in Biotechnol.* **8**, 305–333.

Moses C. O., Nordstrom D. K., Herman J. S., and Mills A. L. (1987) Aqueous pyrite oxidation by dissolved-oxygen and ferric iron. *Geochim. Cosmochim. Acta* **51**, 1561–1571.

National Research Council. (1993) *In Situ Bioremediation; When Does it Work?* National Academy of Sciences, National Academy Press, Washington, D.C.

National Research Council. (1994a) *Ranking Hazardous Waste Sites.* National Academy of Sciences, National Academy Press, Washington, D.C.

National Research Council. (1994b) *Alternatives for Groundwater Cleanup.* National Academy of Sciences, National Academy Press, Washington, D.C.

Nelson M. J. K., Montgomery S. O., and Pritchard P. H. (1988) Trichloroethylene metabolism by microorganisms that degrade aromatic compounds. *Appl. Environ. Microbiol.* **54**, 604–606.

Newell C. J., McLeod K. R., and Gonzales J. R. (1996) BIOSCREEN Natural Attenuation Decision Support System: User's Manual Version 1.3.

Ney R. E. J. (1995) *Fate and Transport of Organic Chemicals.* Government Institutes.

Nielsen P. H., Bierge P. L., Nielsen P., Smith P., and Christensen T. H. (1996) *In situ* and laboratory determined first-order degradation rate constants of specific organic compounds in an aerobic aquifer. *Environ. Sci. Technol.* **30**, 31–37.

Nuclear Energy Agency. (1981) *The Environmental and Biological Behaviour of Plutonium and some other Transuranium Elements.*

Oelkers E., Steefel C. I., and Lichtner P. (1996) *Reactive Transport in Porous Media.* Mineralogical Society of America, Washington, D.C.

Ohnuki T. (1994) Sorption characteristics of cesium on sandy soils and their components. *Radiochim. Acta* **65**, 75–80.

Olson D. (1996) Intrinsic PCB remediation in sediments. *IBC's Second Annual International Symposium on Intrinsic Bioremediation.*

Otfjord G. D., Puhakka J. A., and Ferguson J. F. (1994) Reductive dechlorination of Aroclor 1254 by marine sediment cultures. *Environ. Sci. Technol.* **28**, 2286–2294.

Oudot J., Ambles A., Bourgeois S., Gatellier C., and Sebyera N. (1989) Hydrocarbon infiltration and biodegradation in a landfarming experiment. *Env. Poll.* **59**(17–40).

Palmer C. D. and Puls R. W.-G. W. I. (1994) Natural attenuation of hexavalent chromium in groundwater and soils. *Ground Water Issue.*

Park C. K., Woo S. I., Tanaka T., and Kamiyama H. (1992) Sorption and desorption behavior of 60Co, and 137Cs in a porous tuff: Mechanisms and Kinetics. *J. Nuc. Sci. Technol.* **29**, 1184–1193.

Payne T. E., Davis J. A., and Waite T. D. (1994) Uranium retention by weathered schists — the role of iron minerals. *Radiochim. Acta* **66/67**, 297–303.

Pennington J. C., Hayes C. A., Myers K. F., Ochman M., Gunnison D., Felt D. R., and McCormick E. F. (1995) Fate of 2,4,6-Trinitrotoluene in a simulated compost system. *Chemosphere* **30**, 429–438.

Perdue E. M. (1983) Association of organic pollutants with humic substances: Partitioning equilibria and hydrolysis kinetics. In *Aquatic and Terrestrial Materials* (R. F. Chritman, Ed.), pp. 441–460. Ann Arbor Press, Ann Arbor, MI.

Peterson N. M. (1983) 1983 survey of landfills. *Waste Age* **March 1983**, 37–40.

Pickens J. F. and Grisak G. E. (1981) Scale-dependent disperion in a stratified granular aquifer. *J. Water. Resourc. Res.* **17**, 1191–1211.

Plummer L. N., Jones B. F., and Truesdell A. H. (1976) WATEQF-A Fortran IV version of WATEQ, a computer program for calculating chemical equilibrium of natural waters. *U.S. Geol. Survey Water Resources Investigations*, 76-13.

Portillo R. (1992) Mill tailings remediation. In *Deserts as Dumps?* (C. C. Reith and R. M. Thomson, Eds.), pp. 281–302. UNM Press, Albuquerque, NM.

Probstein R. F. and Hicks R. E. (1993) Removal of contaminants from soils by electric fields. *Science* **260**, 498–503.

Puls R. W. (1995) Natural Attenuation of Hexavalent Chromium. *Ground Water Currents* **July**(12).

Rabus R., Nordhaus R., Ludwig W., and Widdel F. (1993) Complete oxidation of toluene under strictly anoxic conditions by a new sulfate-reducing bacterium. *Appl. Environ. Microbiol* **59**, 1444–1451.

Rai, D. and Zachara, J. M. (1984) *Chemical Attenuation Rates, Coefficients, and Constants in Leachate Migration*, Electric Power Research Institute, Palo Alto.

Rautman C. A. (1996) Geostatistics and cost-effective environmental remediation. *Fifth International Geostatistics Congress.*

Raymond R. L., Hudson J. O., and Jamison J. O. (1976) Oil degradation in soil. *Appl. Env. Microbiol.* **31**, 522–535.

Reinhard M., Barker J. F., and Goodman N. L. (1984) Occurrence and distribution of organic chemicals in two landfill leachate plumes. *Environ. Sci. Technol.* **18**, 953–961.

Renner R. (1995) When is lead a health risk? *Environ. Sci. Technol.* **29**, 256A–261A.

Revesz K., Coplen T. B., Baedecker M. J., Glynn P. D., and Hult M. (1995) Methane production and consumption monitored by stable H and C isotope ratios at a crude oil spill site, Bemidji, Minnesota. *Applied Geochemistry* **10**, 505–516.

Rice D. W., Dooher B. P., Cullen S. J., Everett L. G., Kastenberg W. E., Grose R. D., and Marino M. A. (1995) Recommendations to improve the cleanup process for California's leaking underground fuel tanks. *Lawrence Livermore National Labratory Report* **UCRL-AR-121762**.

Rieuwerts J. S. and Farago M. E. (1995) Lead contamination in smelting and mining environments and variations in chemical forms and bioavailability. *Chemical Speciation and Bioavailability* **7**, 113–123.

Rifai H. S., Bendient P. B., Wilson J. T., Miller K. M., and Armstrong J. M. (1988) Biodegradation modeling at aviation fuel spill site. *J. Env. Eng.* **114**, 1007–1029.

Rifai H. S., Bordon R. C., and Wilson J. T. (1995a) Intrinsic bioattenuation for subsurface restoration. In *Intrinsic Biodegradation* (R. E. Hinchee, J. T. Wilson, and D. C. Downey, Eds.). Battelle Press, Columbus, OH.

Rifai H. S., Newell C. J., Miller R. N., Taffinder S., and Rounsaville M. (1995b) Simulation of natural attenuation with multiple electron acceptors. In *Intrinsic Biodegradation* (R. E. Hinchee, J. T. Wilson, and D. C. Downey, Eds.), pp. 53–56. Battelle Press, Columbus, OH.

Riley R. G., Zachara J. M., and Wobber F. J. (1992) Chemical contaminants on DOE lands and selection of contaminant mixtures for subsurface science research. US-DOE.

Rittman B. and McCarty P. L. (1980) Model of steady-state biofilm kinetics. *Biotech. Bioeng.* **22**, 2343–2357.

Ritz S. M. (1996) Remediation by natural attenuation: A state policy overview. *IBC's Second Annual International Symposium on Intrinsic Bioremediation.*

Ruby M. V., Davis A., and Nicholson A. (1994) *In situ* formation of lead phosphates in soils as a method to immobilize lead. *Environ. Sci. Technol.* **28**, 646–654.

Ruby M. V., Davis A., Schoof R., Eberle S., and Sellstone C. M. (1996) Estimation of lead and arsenic bioavailability using a physiologically based extraction test. *Environ. Sci. Technol.* **30**, 422–430.

Rugge K., Bjerg P., and Christensen T. H. (1995) Distribution of organic compounds from municipal solid waste in the groundwater downgradient of a landfill (Grindsted, Denmark). *Environ. Sci. Technol.* **29**, 1395–1400.

Runnels D. L. and Larson J. L. (1986) A laboratory study of electrmigration as a possible field technique for removal of contaminants from groundwater. *Groundwater Monitoring Rev.*(Summer), 88–91.

Russell M., Colglazier E. W., and English M. R. (1991) Hazardous Waste Remediation: The Task Ahead. *Waste Management Research and Education Institute, University of Tennessee.*

SAIC-GeoSafe Corporation. (1995) *In Situ* Vitrification (ISV) Technology: Innovative Technology Evaluation Report.

Salanitro J. P. (1993) The role of bioattenuation in the management of aromatic hydrocarbon plumes in aquifers. *Ground Water Monitoring and Remediation* **13**, 150–161.

Schnoor J. L., Licht L. A., McCutcheon S. C., and Carreira L. H. (1995) Phytoremediation of organic and nutrient contaminants. *Environ. Sci. Technol.* **29**, A318–A323.

Schocher R. J., Seyfried B., Vasquez F., and Zeyer J. (1991) Anaerobic degradation of toluene by pure cultures of denitrifying bacteria. *J. Arch. Microbiol.* **157**, 7–12.

Schock M. R. (1980) Response of lead solubility to dissolved carbonate in drinking water. *J. Am. Water Works Assoc.*(72), 695–704.

Schwarzenbach R. P., Gschwend P. M., and Imboden D. M. (1993) *Environmental Organic Chemistry.* Wiley-Interscience, New York.

Scott M. J. and Morgan J. J. (1990) Energetics of conservative properties of redox systems. In *Chemical Modeling of Aqueous Systems II* (D. Melchior and R. Bassett, Eds.).

Shelley M. D., Autentrieth R. L., Wild J. R., and Dale B. E. (1996) Thermodynamic analysis of trinitrotoluene biodegradation and mineralization pathways. *Biotech. and Bioeng.* **51**, 198–205.

Siegel D. I., Stoner D., Byrnes T., and Bennett P. C. (1992) A Geochemical process approach to identify inorganic and organic groundwater contaminants, 1291–1301. *Ground Water Management* (2), National Water Resources Association.

Sims R., Lawless T. A., Alexander J. A., Bennett D. G., and Read D. (1996) Uranium migration through intact sandstone: effect of pollutant concentration and the reversibility of uptake. *J. Contam. Hydrol.* **21**, 215–228.

Smith L. A., Means J. L., Chen A., Alleman B., Chapman C. C., Tixier J. S. J., Brauning S. E., Gavaskar A., and Royer M. D. (1995) *Remedial Options for Metals-Contaminated Sites.* CRC Press, Boca Raton, FL.

Spain J. C. (1995) Biodegradation of nitroaromatic compounds. *Ann. Rev. Microbiol.* **49**, 523–555.

Stone A. T. (1987) Reductive dissolution of manganese (III/IV) oxides by substituted phenols. *Environ. Sci. Technol.* **21**, 979–988.

Stumm W. (1995) The inner-sphere surface complex: a key to understanding surface reactivity. *Advances in Chemistry Ser.* **244**, 1–32.

Stumm W. and Morgan J. J. (1981) *Aquatic Chemistry.* Wiley-Interscience, New York.

Stumm W. and Morgan J. J. (1996) *Aquatic Chemistry.* Wiley-Interscience, New York.

Taylor S. W., Milly P. C. D., and Jaffe P. R. (1990) Biofilm growth and the related changes in the physical properties of porous medium; permeability. *Water Resourc. Res.* **26**, 2161–2169.

Terauds V. (1996) Making or breaking the case for natural attenuation. *IBC's Second Annual International Symposium on Intrinsic Bioremediation.*

Thierrin J., Davis G. B., Berber C., Patterson B. M., Pribac F., Power T. R., and Lambert M. (1993) Natural degradation rates of BTEX compounds and napthalene in a sulfate reducing groundwater environment. *Hydrological Sciences J.* **38**, 309–322.

Thomas J. M. and Ward C. H. (1989) *In situ* biorestoration of organic contaminants in the subsurface. *Environ. Sci. Technol.* **23**, 760–766.

Toze S. G., Power T. R., and Davis G. B. (1995) Relating BTEX degradation to the biogeochemistry of an aerobic aquifer. In *Intrinsic Bioremediation* (R. E. Hinchee, J. T. Wilson, and D. C. Downey, Eds.). Battelle Press, Columsus, OH.

Trapp S. and MacFarlane J. C. (1995) *Plant Contamination: Modeling and Simulation of Organic Chemical Processes.* Lewis Publishers/CRC Press, Boca Raton, FL.

Tratnyek P. G. (1996) Putting corrosion to use: remediating contaminated groundwater with zero-valent metals. *Chemistry & Industry*(1 July), 499–503.

Travis C. C. and Doty C. B. (1990) Can contaminated aquifers at Superfund sites be remediated? *Environ. Sci. Technol.* **24**, 1464–1466.

Trust B. A., Mueller J. G., Coffin R. B., and Cifuentes L. A. (1995) The Biodegradation of fluoranthene as monitored using stable carbon isotopes. In *Monitoring and Verification of Bioremediation* (R. E. Hinchee, G. S. Douglas, and S. K. Ong, Eds.). Battelle Press, Columbus, OH.

U.S. Bureau of Mines. (1972) *Minerals yearbook 1970.* U.S. Bureau of Mines.

U.S. Environmental Protection Agency. (1990) Characterization of municipal solid waste in the United States: 1990 update.

U.S. Environmental Protection Agency. (1986) *Underground motor fuel storage tanks: A national survey.*

U.S. Environmental Protection Agency. (1993) Cleaning up the Nation's Waste Sites: Markets and Technology Trends. USEPA, Office of Solid Waste and Emergency Response.

U.S. Environmental Protection Agency. (1983) *Methods for chemical analysis of water and wastes.*

U.S. Environmental Protection Agency. (1986) *Test methods for evaluating solid waste, physical and chemical methods, 3rd ed. SW-486.*

Valsami-Jones E., Ragnarsdottir K. V., Mann T., and Kemp A. J. (1995) An experimental investigation of the potential of apatite as a radioactive and industrial waste scavenger. (Personal communication).

Van de Velde K. D., Marley M. C., Studer J., and Wagner D. M. (1995) Stable carbon isotope analysis to verify bioremediation and bioattenuation. In *Monitoring and Verification of Bioremediation* (R. E. Hinchee, G. S. Douglas, and S. K. Ong, Eds.), pp. 241–257. Battelle Press, Columbus, OH.

Vanwarmerdam E. M., Frape S. K., Aravena R., Drimmie R. J., Flatt H., and Cherry J. A. (1995) Stable chlorine and carbon isotope measurements of selected chlorinated organic solvents. *Appl. Geochem.* **10**, 547–552.

Vogel T. M., Criddle C. S., and McCarthy P. L. (1987) Transformations of halogenated aliphatic compounds. *Environ. Sci. Technol.* **21**(722–736).

Vogel T. M. and Grbic-Galic D. (1986) Incorporation of oxygen from water into toluene and benzene during anaerobic fermentative transformation. *Appl. Environ. Microbiol.* **52**, 200–202.

Vogel T. M. and McCarty P. L. (1985) Biotransformation of tetrachloroethylene to trichloroethylene, dichloroethylene, vinyl chloride, and carbon dioxide under methanogenic conditions. *Applied and Environ. Microbiology* **49**, 1080–1083.

Wachtershauser G. (1988) Before enzymes and templates: Theory of surface metabolism. *Microbiol. Rev.* **Dec. 1988**, 452–484.

Wang E. X., Bormann F. H., and Benoit G. (1995) Evidence of complete retention of atmospheric lead in the soils of northern hardwood forested ecosystems. *Environ. Sci. Technol.* **29**, 735–739.

Weaver J. W., Wilson J. T., and Kampbell D. H. (1996) Extraction of degradation rate constants from the St. Joseph, Michigan trichloroethylene site. *Symposium on Natural Attenuation of Chlorinated Organics in Ground Water.*

Webb E. K., Conrad S., and Breeden R. (1993) A probabilistic approach to site characterization for the Superfund program. *Federal Environmental Restoration Conference.*

Welch S. A. and Ullman W. J. (1993) The effect of organic acids on plagioclase dissolution rates and stoichiometry. *Geochim. Cosmochim. Acta* **57**, 2725–2736.

White A. F., Delany J. M., Narasimhan T. N., and Smith A. (1984) Groundwater contamination from an inactive uranium mill tailings pile 1. Application of a chemical mixing model. *Water Resourc. Res.* **20**, 1743–1752.

Wiedemeier T. H., Swanson M. A., Moutoux D. E., Wilson J. T., Kampbell D. H., Hansen J. E., and Haas P. (1996) Overview of the technical protocol for natural attenuation of chlorinated aliphatic hydrocarbons in groundwater under development for the U.S. Air Force Center for Environmental Excellence. *EPA Symposium on Natural Attenuation of Chlorinated Solvents.*

Wiedemeier T. H., Swanson M. A., Wilson J. T., Kampbell D. H., Miller R. N., and Hansen J. E. (1995a) Patterns of intrinsic bioremediation at two U.S. Airforce Bases. In *Intrinsic Bioremediation* (R. E. Hinchee, J. T. Wilson, and D. C. Downey, Eds.), pp. 31–52. Battelle Press, Columbus, OH.

Wiedemeier T. H., Wilson J. T., Kampbell D. H., Miller R. N., and Hansen J. E. (1995b) Technical protocol for implementing intrinsic remediation with long-term monitoring for natural attenuation of fuel contaminent dissolved in groundwater. *Air Force Center for Technical Excellence, Technology Transfer Division* **1 & 2**.

Wierenga P. J. (1995) Water and solute transport and storage. In *Handbook of Vadose Zone Characterization and Monitoring*, pp. 41–60. Lewis Publishers/CRC Press, Boca Raton, FL.

Wilson B. H., Wilson J. T., Kampbell D. H., Bledsoe B. E., and Armstrong J. M. (1990) Biotransformation of monoaromatic and chlorinated hydrocarbons at an aviation gasoline spill site. *Geomicrobiol. J.* **8**, 225–240.

Wilson J. T., McNabb J. F., Cochran J. W., Wang T. H., Tomson M. B., and Bedient P. B. (1985) Influence of microbial adaptation on the fate of organic pollutants in groundwater. *Environ. Toxicol. and Chem.* **4**, 721–726.

Wilson J. T., Pfeffer F. M., Weaver J. W., Kampbell D. H., Wiedemeier T. H., Miller R. N., and Hansen J. E. (1994a) Intrinsic bioremediation of JP-4 jet fuel. *EPA Symposium on Intrinsic Bioremediation of Ground Water.*

Wilson J. T., Weaver J. W., and Kampbell D. H. (1994b) Intrinsic bioremediation of TCE in groundwater at an NPL site in St. Joseph, Michigan. *EPA Symposium on Natural Attenuation of Chlorinated Solvents.*

Wolery T. J. (1983) A computer program for geochemical aqueous speciation-solubility calculations: user's guide and documentation. *Lawence Livermore National Laboratory Report* **UCRL-53414**.

Xu M. and Eckstein Y. (1995) Use of weighted least-squares method in evaluation of the relationship between dispersivity and scale. *J. of Ground Water* **33**, 905–908.

Yong R. N., Galvez-Cloutier R., and Phadungchewit Y. (1993) Selective sequential extraction analysis of heavy-metal retention in soil. *Can. Geotech. J.* **30**, 834–847.

Zachara J. M., Cowan C. E., and Resch C. T. (1991) Sorption of divalent metals on calcite. *Geochim. Cosmochim. Acta* **55**, 1549–1563.

Index

A

Acceptable daily intake (ADI), 159
Acceptable risk, 158
Acetate, 112, 148, 151, 152
Acidic solutions, 45
Acid mine drainage, 29, 118–120
Adsorption (sorption), 3, 52–54, 61–65, 78
  dead-end pores and, 148
  demonstrating metal natural attenuation, 146–148, 151–155
  irreversible sorption of metals, 95–100
  isotherms, 63–64
  K_d and, 95, 98–99, 147–158
  metals, case studies, 95–106
    cesium, 96–97
    lead, 192
    radioactive soils, 103–105
    sorption to carbonates, 97–98
    sorption to iron (hydr)oxides, 100–102
    uranium, 98–100
    uranium mill tailings, 105–106
  mineral growth and, 53–54
  NAPLs to mineral surfaces, 35–36
  pH and, 53–54
  pump-and-treat problems, 64, 67
  radionuclides, 23
  retardation factor approach, 61
  transport and, 61–64, See also Desorption
Advection, 36
Aerobic degradation, 68
Agrico Chemical Co. site, FL, 213–214
Air Force Center for Environmental Excellence (AFCEE), 78
  chlorinated solvents protocol, 135–146
  fuel hydrocarbons protocol, 121–134, 165
Air sparging, 12, 81
Alcoa (Vancouver smelter) site, WA, 206–210
Aldrin, 148
Alkaline solutions, 45
Alkalinity maps, 123
Alkane biodegradation pathways, 70
Alternatives analysis, 10–11, 167–169
Aluminum, 106
  NPL sites with natural attenuation as part of final remedy, 202
  production site, 206–210
  solubility, 49

American Society for Testing and Materials (ASTM), 162, 164, 219
Americium, 22, 38, 104, 108
  carbonate mineral phases, 47
  data needs for natural attenuation, 154
  likely natural attenuation pathways, 153
  retardation factors in repository rocks, 103
Ammonia, 199
Ammonium ion oxidation, 68
Anaerobic degradation, 68, 70–71
Anoxic limestone drains, 119
Antimony, 18, 199–200
Aquifer
  defined, 32
  remediation, 1, 168
Archaebacteria, 68
Aroclor, 188
Arsenic, 21
  bioavailability, 118
  data needs for natural attenuation, 154
  drinking water standards, 18
  hazardous waste sources, 16
  likely natural attenuation pathways, 153
  NPL sites with natural attenuation as part of final remedy, 188, 193, 195, 196, 213, 214
Arsines, 21
Asphalt-hardened cap, 87
ASTM, 162, 164, 219

B

B & B Chemical Company site, FL, 200
Barium, 105
  data needs for natural attenuation, 154
  drinking water standards, 18
  hazardous waste sources, 16, 20–21
  likely natural attenuation pathways, 153
Basalt, 100–103
Bendix Corporation/Allied Automotive site, MI, 215–216
Benzene, 27, 76
  biodegradation, 58, 66, 68, 69
  drinking water standards, 18, 26
  hazardous waste sources, 16
  NPL sites with natural attenuation as part of final remedy, 183, 185, 188, 193, 196, 199, 211, 212
  vapor characteristics, 45, 47

235

Beryllium, 18, 21
Bio-sparging, 12
Bioavailability
  dead-end pores and, 148
  lead and arsenic, 118
  metal mineral forms, 47–48
Biodegradation, 65
  BIOSCREEN, 78, 129–133, 142
  calculating in the field, 123, 128–129
  carbon isotope analysis, 59–60
  chlorinated organics, 28, 71–73
  constants for chlorinated solvents, 142
  contaminant susceptibility, 74
  electron acceptors, 68–70, 78
  enzymes, 70
  general chemistry, 70–73
  half-lives, 4
  in situ bioremediation, 81, 85
  RBCA approach, 164
  rates, 73–77
  recent studies indicating, 75
  site- and contaminant-specific data, 124–127
  TNT, 114
Bioplume II, 130
BIOSCREEN, 78, 129–133, 142
Bioventing, 12, 81
Bromide tracers, 40
BTEX compounds, 31, 58, 60, 114
  biodegradation, 68, 70, 75, 78
  demonstrating natural attenuation, 143

C

Cadmium, 78, 165
  data needs for natural attenuation, 154
  drinking water standards, 18
  hazardous waste sources, 16–18
  likely natural attenuation pathways, 153
  NPL sites with natural attenuation as part of final remedy, 186, 193, 214
  solubility, 49
  sorption, 47, 97, 150
Calcite extraction, 152
Calcium carbonate
  acid mine drainage remediation, 119–120
  extraction agents, 152
  mineral formation, 47
California Environmental Protection Agency, 7, 12
Cancer risk, 159
Capillary forces, 34
Capping systems, 85–86
  for radon, 106
Carbon-13/Carbon-12 ratio, 58–60, 87
Carbon-14, 58, 108
Carbonates, See also Calcium carbonate
  dissolution rates, 55
  metal sorption, 47, 97–98
  soil amendments, 119–120
Carbon dioxide
  electron acceptor utilization sequence, 68, 70
  protocol for demonstrating natural attenuation, 139
  soil gases, 45–46
  stable carbon isotope analysis, 59
Carbon isotope analysis, 58–60, 87
Carbonic acid, 44
Carbon tetrachloride, 66, 76
  biodegradation pathways, 72, 73
  drinking water standards, 18
  NPL sites with natural attenuation as part of final remedy, 214
Carcinogenic contaminants, 4
Carson, Rachel, 3
Cement, radioactive waste burial in, 109
CERCLA, 1, 157, See also National Priority List (NPL) sites
  cleanup goals/standards, 167–168
  cleanup standards, 9–11
  history, 7–8
  liability scheme, 1, 8, 10, 157
  mechanics, 8–11
  natural attenuation and, 11–12
  seeds of failure, 13
  time to implementation, 4
  ultimate government liability, 157
  unintended beneficiaries, 157
  unreasonable premise of immortal toxicity, 171
Cesium, 3, 22, 24, 96–97, 108
  data needs for natural attenuation, 154
  likely natural attenuation pathways, 154
  retardation factors in repository rocks, 103
CFC-113, 75
Chelating agents, 47, 82, 105
Chemical attenuation, 43–65
  adsorption, 52–54, 61–65, See Adsorption
  contaminant concentration, 43–44
  electron transfer reactions, 50–52
  equilibrium constant, 43–44
  gas dissolution/exsolution, 45–47
  geochemical modeling, 56–58
  ion pairing, 44–45
  isotopic modeling, 58–61
  metal hydroxide solubilities, 48–50
  mineral dissolution and contaminant availability, 54–56
  mineral growth, 47–50
  reversible and irreversible reactions, 43
Chemical compound information resource, 219
Chemical treatments, 86
Chemical warfare agents, 21

Index

Chlordane, 148
Chloride
  NPL sites with natural attenuation as part of final remedy, 199
  protocol for demonstrating natural attenuation, 141, 143, 145
  tracers, 40
Chlorinated hydrocarbons, 27–28, See also specific chemicals
  biodegradation pathways, 71–73
  demonstrating natural attenuation, 135–146, See also under Natural attenuation demonstration for regulatory approval
  natural attenuation case studies, 112–114
  significant biodegradation, 75
Chlorine isotopes, 60–61
Chlorobenzene, 185, 193, 200
Chloroethane, 144
Chloroform, 4, 16, 66, 191
Chromium, 78
  chemical reduction, 86
  drinking water standards, 18
  hazardous waste sources, 16, 20
  NPL sites with natural attenuation as part of final remedy, 189, 201, 214, 216
Chromium(III), 116
Chromium(VI) (CrO_4^{2-}), 115–117, 165, 214
  data needs for natural attenuation, 154
  likely natural attenuation pathways, 153
Citrate-dithionate solution, 152
Citrate buffers, 148, 151
Clay
  capping systems, 85–86, 191
  hydraulic conductivity, 32
  impermeable layers, 82
  radionuclide retardation factors, 103
  sorption, 53, 54, 96–97
Cleanup goals and standards, 1, 9–11, 167–169, See also Drinking water standards
Coakley Landfill site, NH, 196–197
Coal combustion, 19
Coal tar production site, 117–118
Cobalt, 22, 105, 165
  data needs for natural attenuation, 154
  likely natural attenuation pathways, 154
  sorption, 104
Colloidal transport, 37–38
Composting of TNT, 114
Comprehensive Environmental Response, Compensation and Liability Act, See CERCLA
Computer applications, 78, 129–133
  geochemical modeling, 56–57
  WWW resources, 217–219

Conductivity, protocol for demonstrating natural attenuation, 141
Constructed wetlands, 86
Contaminant contour maps, 123
Contaminant mobility, See Desorption; Groundwater flow
Contaminant sources, 15–29, See Hazardous waste sources
Contaminant transport rules-of-thumb, 77–78
Copper, 78
  carbonate mineral phases, 47
  hazardous waste sources, 16
  uranium mill tailings, 106
Costs
  aquifer remediation, 168
  hazardous waste clean-up expenditures, 1, 8
  in situ vitrification, 83
  legal fees, 8
  soil flushing, 82
  soil immobilization, 86
Creosote production site, 117
Cresols, 76
Curium, 22
Cyanide
  drinking water standards, 18
  NPL sites with natural attenuation as part of final remedy, 199, 208–209
  spent potlining, 207

D

Darcy's law, 31–34
Davie Landfill site, FL, 199–200
DDT, 3, 4, 47, 66, 148
  degradation half-life, 75, 76
Dead-end pores, 3, 148
Debye-Huckel law, 44
Degradation rate constant, 129
Dense non-aqueous-phase liquids (DNAPLs), 35, 110–114
Department of Defense (DOD) sites, 7
  fuel hydrocarbon cleanup protocol, 121–134
  Installation Restoration Program (IRP) sites, 189, 210, 215
  NPL sites with natural attenuation as part of final remedy, 188–189, 210–212, 214–215
  quantitative and qualitative risk assessment and prioritization, 160
Department of Energy (DOE) sites, 7
  metal-contaminated sites, 17
  quantitative and qualitative risk assessment and prioritization, 160
  radioactive waste facilities, 23, 24
  WWW resources, 217

Desorption, 3, 61–64, See also Adsorption; K_d
  isotherms, 63–64
  rates for metals, 146–147
  thermal, 82
Desorption-enhancing agents, 82, 105
Detergent tracers, 40
Diamond Shamrock site, GA, 202–203
Dibenzofuran, 117–118
1,2–Dibromomethane, 148
1,2–Dichlorobenzene, 76
Dichloroethane
  drinking water standards, 18
  hazardous waste sources, 16
  NPL sites with natural attenuation as part of final remedy, 185, 190, 193, 202, 205
1,1–Dichloroethane, 16
1,2–Dichloroethane, 16, 18, 193, 202, 205
Dichloroethene, 112–114, 185
  demonstrating natural attenuation, 144, 145
1,1–Dichloroethene, 193
  demonstrating natural attenuation, 144
  hazardous waste sources, 16
1,2–Dichloroethene, 16
Dichloroethylene, 200
  drinking water standards, 18
  NPL sites with natural attenuation as part of final remedy, 190
1,1–Dichloroethylene, 18
1,2–Dichloroethylene, 18
Dichloromethane, 112
Di(2–ethylhexyl)phthalate, 16
Diffusion coefficient, 36–37
Dioxins, 75
Dispersion, 36–37
Dissolution rate, soil minerals, 55
Ditching, 85
Dolomite extraction, 152
Dover Air Force Base site, DE, 214–215
Drinking water standards, 1, 9, 18
Drycleaning facility sites, 168

E

Earthquakes, 109
Ecological models, 219
Ecological risk assessment, 5
Edieldrin, 148
EDTA, 47, 105
Electrokinetics, 82
Electron acceptors, 68–70, 78
  maps, 123
Electron donors, 78
Electron transfer reactions, 50–52
Engineering approaches, 81–87, 157, See also Pump-and-treat systems; Remediation
  chemical treatments, 86
  DNAPLs and, 110, 112
  ineffectiveness of, 2
  in situ bioremediation, 81, 85
  in situ extraction technologies, 81–83
  in situ vitrification, 83
  physical containment, 83, 85
  phytoremediation, 83
  reactive barriers, 83
  regulatory biases, 11, 13
  soil stabilization/immobilization, 85–86
Environmental Protection Agency (EPA), 1, See also National Priorities List (NPL) sites
  bias toward engineered solutions, 11, 13
  cleanup supervision, 8
  natural attenuation policy, 169
  quantitative and qualitative risk assessment and prioritization, 160
  WWW resources, 217
Environmental risks, 159
Enzymatic biodegradation, 70
EPA, See Environmental Protection Agency
EQ3/6, 56
Equilibrium constant, 43–44
Ethane, 140, 144
Ethene, 52, 140, 144
Ethylbenzene, 27
  drinking water standards, 18, 26
  hazardous waste sources, 16
  NPL sites with natural attenuation as part of final remedy, 183
European environmental information, 219
Explosives contamination, 114–115
Exposure assessments, exaggeration of contaminant risks, 9–10
Exposure pathways analysis, 131
Extraction agents, 152

F

Federal facility sites, See Department of Defense (DOD) sites; Department of Energy (DOE) sites
Fick's first law, 36
Fish, 183, 186, 188
Fluid flow velocity, 31, 77
Fluid transport, See Groundwater flow
Fluoride, 18, 208–209, 213
Flushing, See Soil flushing
Freon, 75
Freundlich model, 62–63
Fuel rods, 109
Fuel tanks, See Leaking underground fuel tanks
Fulvic acids, 53

Index

G

Galena, 120
Gas dissolution/exsolution, 45–47
Gas monitoring system, 191
Gasoline station tanks, See Leaking underground fuel tanks
Gas vents, 193
Geochemical modeling, 56–58
Geochemistry, See Chemical attenuation
Geologic disposal of radioactive waste, 109
Geomembrane curtains, 86
Glossary, 179–181
Grading, 85
Granite, 103
Granulated activated carbon filters (GACs), 168
Grazing, 104
Groundwater flow, 31–42
 colloidal transport, 37–38
 Darcy's law and hydraulic conductivity, 31–34
 diffusion and dispersion, 36–37
 geochemical modeling, 58
 inorganic species, 46
 nonaqueous phase liquids (NAPLs), 35–36
 reactions with aquifer solids, 58
 tracers, 38–40
 unsaturated flows, 34–36
Groundwater interceptor trench, 198
Grout curtains, 86

H

Half-life
 contaminant degradation, 23, 73–77
 radioisotopes, 3, 4, 23, 108
Halogenated hydrocarbons, 27–28, See also Chlorinated hydrocarbons
Hazardous waste sources, 15–29
 arsenic, 21
 landfill leachates, 29
 metals, 15–21
  barium, 20–21
  beryllium, 21
  cadmium, 17–18
  chromium, 20
  lead, 15, 18–19
  mercury, 19–20
  nickel, 21
  silver, 21
  zinc, 20
 mine drainage, 29
 nitrate, 21
 organics, 25–29
  halogenated hydrocarbons, 27–28
  nitroaromatics, 26–27

 PCBs, 28
 petroleum hydrocarbons, 26
 petroleum production wastes, 28–29
 radioactive waste, 22–25
Health risks, 1, 158
 contaminant quantity and, 15–16
 DDT model, 3
 false premises of current beliefs, 2–3
 quantification, 158–160
 site prioritization, 160–161
  RBCAs, 161–164
 sources, See Hazardous waste sources
 unrealistic assumptions exaggerating calculated risk, 9–10, 159–160
Heating, 82
Henry's law, 45
Heptachlor, 148
High level radioactive waste, 22, 109–110, See also Radionuclides
Horizontal barriers, 87
Hospital wastes, 15, 22
Household waste, 15
Humic substances, 53
Humin, 53
Hydraulic conductivity, 31–34
 in situ bioremediation, 85
Hydrogen fluoride, 152
Hydrogen gas, as system redox monitor, 52, 53
Hydrogen peroxide, 152
Hydrolysis, 68

I

Incineration, 187, 199
In situ bioremediation, 81, 85
In situ extraction technologies, 81–83
In situ vitrification, 83
Installation Restoration Program (IRP), 189, 210, 215
Internet resources, 217–219
Intrinsic remediation, See Natural attenuation
Iodine
 data needs for natural attenuation, 154
 likely natural attenuation pathways, 154
 retardation factors in repository rocks, 103
Iodine-129, 108, 109
Iodine-131, 24
Ion pairing, 44–45
Iron
 electron transfer reactions, 50
 NPL sites with natural attenuation as part of final remedy, 202
 solubility, 48–50
 spent potlining, 207
 uranium mill tailings, 106

Iron(II), 50–51, 87
  protocol for demonstrating natural attenuation, 140, 143
Iron(III), 48, 50–51, See also Iron (hydr)oxides
  chlorinated hydrocarbon degradation, 114
  electron acceptor utilization sequence, 68, 69–70
Iron (hydr)oxides, 78
  acid mine drainage neutralization, 29
  electron acceptor utilization sequence, 69
  extraction agents, 152
  mineral growth reactions, 48–50
  sorption to, 53–54, 100–102
Iron ore mines, 194–196
Irreversible sorption, 95–96, See also Adsorption
  cesium onto clay, 96–97
  demonstrating for metals, 146–148, 151
  uranium to soils, 98–100
Isotherms, 63
Isotopic modeling, 58–61

J

Jackson Township Landfill site, NJ, 203–4
Jet fuel, 26
Juncos Landfill site, PR, 191–192

K

K_d, 61–63
  field measurement, 148, 155
  radioactive contaminants in soils, 103
  sorption and, 66–67, 95, 98–99, 147–148
Kepone, 148
Ketones, 76, 196, 201
Kin-Buc Landfill site, NJ, 186–187
K_{OW}, See Octanol-water partitioning coefficient

L

Landfill gas monitoring system, 191
Landfill leachates, 29
Landfills, NPL sites with natural attenuation as part of final remedy, 186–187, 190–200, 202–204, 212–213
Langmuir model, 62–63
LASAGNA, 82
Lawrence Livermore National Laboratory, 12
Leachates, 29
Lead, 78, 102, 165
  bioavailability, 118
  carbonate mineral phases, 47
  colloidal transport, 38
  data needs for natural attenuation, 154
  hazardous waste sources 15, 16, 18–19
  iron hydroxide sequestering, 100–102
  likely natural attenuation pathways, 153
  NPL sites with natural attenuation as part of final remedy, 188, 189, 193, 195, 196, 199, 213, 216
  retardation factors in repository rocks, 103
  soil carrying capacity, 47
  solubility, 49
Leaded gasoline, 102
Lead phosphates, 120
Lead pipe solder, 102
Leaking underground fuel tanks (LUFTs), 15
  downloadable LLNL report, 218
  Lawrence Livermore report, 12
  RBCA approach, 162
  state natural attenuation policies, 173–175
Legal costs, 8
Levee system, 184
Limestone, acid mine drainage remediation, 29, 119–120
Longitudinal dispersion, 37
Long-term monitoring, 121, 133, 155
Low level radioactive wastes, 22–23, 108–109
Low permeability zones, 82
LUFTs, See Leaking underground fuel tanks

M

Manganese, 68, 70, 87
  carbonate sorption, 97
  hazardous waste sources, 16
  NPL sites with natural attenuation as part of final remedy, 202
Manufactured-gas plant site, 117
Matrix suction, 34
Maximum Contaminant level (MCL) goals, 1, 9–10
McChord Air Force Base washrack treatment area site, WA, 210–212
Medical wastes, 15, 22
Mercury, 150, 165
  data needs for natural attenuation, 154
  drinking water standards, 18
  hazardous waste sources, 19–20
  likely natural attenuation pathways, 153
  NPL sites with natural attenuation as part of final remedy, 191, 214
Metabolic by-product maps, 123
Metals, 3, See also specific metals
  adsorption and desorption isotherms, 63–64
  carbonate minerals, 47, 97–98
  demonstrating natural attenuation, suggested protocol, 151–155
  data needs, 154

likely natural attenuation pathways, 153–154
soil organic matter, 155
DOE sites, 17
hazardous waste sources, 15–21
natural attenuation case studies, 95–106
RBCAs, 164–165
sequential extraction analysis, 152
soil carrying capacity, 47
sorption, See Adsorption
Metal sulfides, 55, 119, 120
Methane, 60, 68, 140, 145
Methanol, 112
Methyl bromide, 76
Methylene chloride, 16, 188, 201
Methyl ethyl ketone, 76, 196, 201
Methyl tert butyl ether, 40
Microporosity, 3
Mine drainage, 29, 118–120
mineql, 218
Mineral dissolution, 54–56
Mineral growth, 47–50, 53
Mining industry wastes, 15
Mining sites, NPL sites with natural attenuation as part of final remedy, 194–196
MINTEQ, 56
Mobilizing agents, 47, 82
Models, 161, 219
Monroe Township Landfill site, NJ, 192–194
Mosley Road Sanitary Landfill site, OK, 190–191
MTBE, 40, 218

N

Naphthalene, 27, 117–118, 148
NAPLs, See Nonaqueous phase liquids
National Contingency Plan (NCP), 7, 8–11, 169, See also CERCLA
bias toward engineered solutions, 11, 13
cleanup goals, 9–10
national hazardous substance response plan, 9
natural attenuation and, 11
National priority list (NPL) sites, 1, 7
deletion criteria, 206, 210, 211
expansion of, 7
natural attenuation as part of final remediation, 169–170, 183–216
Agrico Chemical Co., FL, 213–214
Alcoa (Vancouver smelter), WA, 206–210
B & B Chemical Company, FL, 200
Bendix Corporation/Allied Automotive, MI, 215–216
Coakley Landfill, NH, 196–197
Davie Landfill, FL, 199–200
Diamond Shamrock, GA, 202–203
Dover Air Force Base, DE, 214–215
Jackson Township Landfill, NJ, 203–204
Juncos Landfill, PR, 191–192
Kin-Buc Landfill, NJ, 186–187
McChord Air Force Base washrack treatment area, WA, 210–212
Monroe Township Landfill, NJ, 192–194
Mosley Road Sanitary Landfill, OK, 190–191
New Castle spill site, DE, 204–206
Oak Grove Sanitary Landfill, MN, 212–213
PSC Resources, MA, 187–188
Ringwood Mines/Landfill, NJ, 194–196
Rockingham Landfill, VT, 197–199
Sheridan, TX, 183–184
Town Garage/Radio Beacon, NH, 189–190
Twin Cities Air Force Reserve, MN, 188–189
Western Sand and Gravel, RI, 184–185
Wilson Concepts, FL, 201
original sites, 9
National Research Council, 11
Natural attenuation, See also Adsorption; Biodegradation
case studies
acid mine drainage, 118–120
bioavailability of lead and arsenic, 118
chlorinated hydrocarbons, 112–114
dense non-aqueous phase liquids, 110–114
landfills, 91–95
metals, 95–96
cesium, 96–97
hexavalent chromium, 115–117
lead, 192
radioactive soils, 103–105
sorption to carbonates, 97–98
sorption to iron (hydr)oxides, 100–102
uranium, 98–100
uranium mill tailings, 105–106
PAHs, 117–118
petroleum hydrocarbons, 87–91
radioactive waste, 108–110
TNT, 114–115
CERCLA and, 11–12, 158
demonstrating, 4–5, See Natural attenuation demonstration for regulatory approval
EPA definition, 2
EPA policy, 169
Lawrence Livermore report, 12
likely pathways for metals, 153–154
NPL sites, See under National Priority List (NPL) sites
quantification of, 4–5
radionuclides, 23–25
state policies, 12, 173–177
surface shielding, 3–4
Natural attenuation demonstration for regulatory approval, 4–5, 121–155

contaminant degradation rates
  BIOSCREEN, 129–133
  field calculations, 123, 128–129
  site- and contaminant-specific data, 124–127
data needs for metals, 154
irreversible sorption for metals, 146–148, 151
long-term monitoring, 121, 133, 155
protocol for chlorinated solvents, 135–146
  analytical protocol, 139–141
  analytical parameters and weighting, 143–144
  biodegradation rate constant, 142
  consistency of predicted and observed plume movement, 142
  documenting biodegradation, 137–141
  site screening, 137, 138
protocol for fuel hydrocarbons, 121–134
  conceptual model, 123
  documenting intrinsic remediation, 123–131
  exposure pathways analysis, 131
  long-term monitoring plan, 133
  negotiation with regulatory agencies, 134
  site characterization, 123
suggested protocol for metals, 151–155
Natural attenuation screening program, 78
Naturally occurring radioactive material (NORM), 29
Natural reactor, 110
Neptunium, 22, 98
New Castle spill site, DE, 204–206
Nickel, 104
  data needs for natural attenuation, 154
  drinking water standards, 18
  hazardous waste sources, 16, 21
  likely natural attenuation pathways, 153
  NPL sites with natural attenuation as part of final remedy, 193
  radioisotope half-lives, 108
  solubility, 49
Nitrate, 21
  drinking water standards, 18
  electron acceptor utilization sequence, 68–69
  likely natural attenuation pathways, 153
  protocol for demonstrating natural attenuation, 140, 143, 145
Nitroaromatic compounds, 26–27, 75
Nonaqueous phase liquids (NAPLs), 35–36, 78
  natural attenuation case studies, 110–114
Nonylphenol, 148
No-project alternative, 169
NPL, See National Priority List (NPL) sites
Nuclear reactor, 109

O

Oak Grove Sanitary Landfill site, MN, 212–213

Octanol-water partitioning coefficient (K_{OW}), 48, 61–62, 65–66
Off-gas treatment system, 83
Oklo Natural reactor, 110
"Old woman who swallowed a fly" analogy, 171
Organic acids, mineral dissolution rates and, 55
Organic hazardous waste sources, 25–29, See also specific substances
Organic matter
  ascribing heavy metals to, 155
  extraction, 152
Oxidation capacity, 78
Oxidation-reduction (redox) reactions, 50–52
Oxidation treatment, 86
Oxygen
  electron acceptor utilization sequence, 68
  protocol for demonstrating natural attenuation, 139, 140, 143, 145
  soil gases, 45–46

P

Paint sludge, 189, 195
PCBs, See Polychlorinated biphenyls
PCE, See Tetrachloroethene
PeeDee Belemnite (PDB), 58–59
Pentachlorophenol, 4
Persistence to distance assessment, 4, 40
Petroleum hydrocarbons, 26, 87–91, See also Leaking underground fuel tanks; specific chemicals
  demonstrating natural attenuation, 121–134, 165, See also under Natural attenuation demonstration for regulatory approval
Petroleum production wastes, 15, 28–29
pH, 45
  digestive solubility and, 118
  lead sorption and, 102
  metal hydroxide solubilities, 48–50
  metal mineral phases and, 47
  metal sorption/precipitation processes, 78
  mineral dissolution rates and, 55, 56
  mineral sorption and, 53–54
  protocol for demonstrating natural attenuation, 141, 143
  uranium mill tailings, 106
Phenanthrene, 27, 148
Phenols, 16, 75
Phosphate, acid mine drainage remediation, 119–120
Physical containment, 83, 85
Phytoremediation, 83, 157
Plutonium, 15, 22, 103–104
  colloidal transport, 38
  data needs for natural attenuation, 154

Index

likely natural attenuation pathways, 153
radioisotope half-life, 108
retardation factors in repository rocks, 103
Point of compliance (POC) well monitoring, 133
Polybrominated biphenyls, 148
Polychlorinated biphenyls (PCBs), 28, 148
  biodegradation, 75
  drinking water standards, 18
  NPL sites with natural attenuation as part of final remedy, 183, 184, 186, 188
  sorption, 148
Polycyclic aromatic hydrocarbons (PAHs), 26
  biodegradation, 70, 75
  isotope fractionation analysis, 60
  natural attenuation case studies, 117–118
  protocol for demonstrating natural attenuation, 139
  solubilities, 65
Porosity, 31, 32
Potassium radioisotopes, 29
Potlining, 207
Probabilistic decision analysis, 160
Procaryotes, 67–68
PSC Resources site, MA, 187–188
Pump-and-treat systems, 11, 13
  adsorption/desorption processes and, 64, 67
  ineffectiveness, 64, 81
Punitive sanctions, 171
Pyrene, 66
Pyrite, 119

R

Radioactive waste, 22–25
  disposal and storage, 108–110
Radionuclides, 3, 78, See also specific radioisotopes
  carbonate mineral phases, 47
  chemical treatment, 86
  half-lives, 3, 4, 108
  hazardous waste sources, 22–25
  likely natural attenuation pathways, 153
  natural attenuation, 23–25, 103–105
  naturally occurring radioactive material (NORM), 29
  Oklo natural reactor, 110
  plant uptake, 104
  problems of field analysis, 151
  retardation factors in repository rocks, 103
  storage, 22–23
Radium, 24, 29, 108
  retardation factors in repository rocks, 103
Radon, 22, 106
RBCAs, 161–164, 177
RCRA, 1

REACT, 56
Reaction transport modeling, 130
Reactive barriers, 83
Redox reactions, 50–52
Reducing agents, 82
Reduction capacity, 78
Reductive dehalogenation, 71
Reference dose, 159
Remediation, See also Engineering approaches; Pump-and-treat systems
  alternatives analysis, 167–169
  cleanup goals and standards, 9–10, 167
  implementing ineffective technologies, 171
  ineffectiveness of engineered extraction, 2
  newer technologies, 12, 81–87, 157
    WWW resources, 219
  numerical standards, 9
  RBCAs, 161–164
  regulatory biases toward engineered solutions, 11, 13
  Superfund Innovative Technology Evaluation (SITE), 216
  technological limits, 11
  total system performance approach, 160–161
  waste stabilization assessment, 86–87
Remediation Technology Development Forum (RTDF), WWW site, 218
Resource Conservation and Recovery Act (RCRA), 1
Retardation factor, 61
  for radionuclides in various repository rocks, 103
Reversible and irreversible chemical reactions, 43
Ringwood Mines/Landfill site, NJ, 194–196
Risk, 158–161, See also Health risks
  acceptable, 158
  assessment, 159
  characterization, 159
Risk-based corrective actions (RBCAs), 161–164, 177
Risk-based screening level (RBSL), 162, 164
Riverton, Wyoming, 106–107
Rock Flats, 104
Rockingham Landfill site, VT, 197–199
R. S. Kerr Environmental Research Center, 78, 130
RTDF (WWW site address), 218
Rusting, 50

S

Safe Driinking Water Act, 9
Salinity
  mineral dissolution rate and, 56
  thermodynamic activities and, 44
Salt, radionuclide retardation factors, 103

Sandy soils, hydraulic conductivity, 32
Saturated zone, 32, 35
Selenium, 18, 98, 106
Sequential extraction analysis, 152
Shale, 103
Sheet pile walls, 86
Shellfish, 186
Sheridan site, TX, 183–184
Silent Spring, 3
Silicates, 87
    extraction agents, 152
Silver, 21
Simazine, 148
SITE, 216
Site screening, RBCA approach, 161–164
Site-specific contaminant target levels (SSTLs), 163
Sludge lagoon, 199
Slurry wall, 86, 187, 214
Smoke detectors, 15
Soil chemistry, See Chemical attenuation
Soil flushing, 82
    adsorption/desorption processes and, 64
    maximum contaminant removal, 96
Soil microorganisms, types of, 67–68
Soil oxidation capacity, 78
Soil stabilization/immobilization, 85–86
Soil vapors and gases, 45–46
    extraction, 12, 81
Solid/fluid distribution, See K_d
Solubility, 48–50
    K_d and, 66–67
    organic chemicals, 65–66
Solubility-enhancing agents, 82
Stable carbon isotope analysis, 58–60, 87
State policies
    acceptance of natural attenuation, 12, 173–177
    RBCA approach for LUFT cleanups, 162
Steam-enhanced extraction, 82
Strontium, 22, 103, 105, 108, 109
    carbonate soil amendments, 120
    data needs for natural attenuation, 154
    likely natural attenuation pathways, 153
    NPL sites with natural attenuation as part of final remedy, 201
Subsurface heterogeneity, 33, 77, 85
Sulfate
    electron acceptor utilization sequence, 68, 70
    NPL sites with natural attenuation as part of final remedy, 199
    protocol for demonstrating natural attenuation, 140, 143, 145
Sulfides, 119, 120, 143
    dissolution rates, 55

NPL sites with natural attenuation as part of final remedy, 199
Superfund Accelerated Cleanup Method (SACM), 198
Superfund Amendments and Reauthorization Act (SARA), 7, See also CERCLA
Superfund factbook, 219
Superfund Innovative Technology Evaluation (SITE), 216
Superfund NPL sites, See National Priority List (NPL) sites
Surface shielding, 3–4
Surfactants, 82

T

TCA, 112, 113
    biodegradation pathways, 72, 73
2,3,7,8–TCDD, 148
TCE, See Trichloroethene
TCLP, 87, 148, 218
Technetium, 24, 98, 108, 109
    data needs for natural attenuation, 154
    likely natural attenuation pathways, 154
    retardation factors in repository rocks, 103
Technological limitations, 11
Temperature, protocol for demonstrating natural attenuation, 141, 143
Terrorists, 109
Tetrachloroethene (PCE), 28
    biodegradation pathways, 72, 73
    degradation half-life, 75, 76
    demonstrating natural attenuation, 143, 145
    drinking water standards, 18
    isotopic analysis, 60–61
    natural attenuation case studies, 112–114
    NPL sites with natural attenuation as part of final remedy, 214
    redox reactions, 51–52
Tetrachloroethylene, 188
Tetraethyllead, 18–19
Tetrahydrofuran, 196
Texas Natural Attenuation Protocol, 219
Thermal desorption, 82
Thermodynamic activities, 44
Thorium, 29, 154
Time bombs, 3
TNT, 26, 27, 114–115
Toluene, 27, 76, 114
    biodegradation model, 74
    drinking water standards, 18, 26
    hazardous waste sources, 16
    NPL sites with natural attenuation as part of final remedy, 183, 185, 202, 212
Tortuosity, 37

Index 245

Total organic carbon (TOC), protocol for demonstrating natural attenuation, 139, 141
Total system performance approach, 160–161
Town Garage/Radio Beacon site, NH, 189–190
Toxicity, See Health risks
Toxicity assessment, unreasonable assumptions leading to overestimation of, 159–160
Toxicity Characteristic Leaching Procedure (TCLP), 86, 148, 218
Tracers, 38–40
Trichloroethane, 185
1,1,1-Trichloroethane, 61
  demonstrating natural attenuation, 144
  drinking water standards, 18
  hazardous waste sources, 16
1,1,2-Trichloroethane, drinking water standards, 18
Trichloroethene (TCE), 28, 47
  biodegradation pathways, 72, 73
  degradation half-life, 75, 76
  demonstrating natural attenuation, 143, 145
  drinking water standards, 18
  isotopic analysis, 60–61
  natural attenuation case studies, 112–114
  NPL sites with natural attenuation as part of final remedy, 202, 205, 214
  redox reactions, 51–52
Trichloroethylene
  hazardous waste sources, 16
  NPL sites with natural attenuation as part of final remedy, 183, 185
Trinitrotoluene (TNT), 26, 27, 114–115
Tris, 205–206
Tritium, 24, 108–109
Tuff, 103
Twin Cities Air Force Reserve site, MN, 188–189

U

Underground fuel tanks, See Leaking underground fuel tanks
Unsaturated (vadose) zone, 32, 34–36
Uranium, 22, 24, 29
  data needs for natural attenuation, 154
  likely natural attenuation pathways, 153
  mill tailings, 22, 105–106
  Oklo natural reactor, 110
  radioisotope half-lives, 108
  retardation factors in repository rocks, 103
  sorption to soils, 98–100, 104, 106

V

Vadose (unsaturated) zone, 32–36
Vanadium, 150
Vapor extraction, 12, 81
Vapor pressure, 47
Vapor transport, 31, 45, 77
Vertical barriers, 85–86
Vertical dispersivity, 37
Vinyl chloride, 51, 76, 113, 114
  biodegradation pathways, 71–72
  demonstrating natural attenuation, 144, 145
  drinking water standards, 18
  hazardous waste sources, 16
  NPL sites with natural attenuation as part of final remedy, 193, 199–200
Vitrification, 83
Volatilization, 47
Volcanic ash, 103
Volcanic eruptions, 109

W

Waste Extraction Test (WET), 86–87
Waste-oil refinery, 187
WATEQ, 56
Water table gradient, 35
Water table topography, 31, 32
Web resources, 217–219
Western Sand and Gravel site, RI, 184–185
WET, 86–87
Wetlands
  constructed, 86
  contamination, 187, 212
Wilson Concepts site, FL, 201
WWW resources, 217–219

X

Xylenes, 27, 76, 117–118
  drinking water standards, 18, 26
  hazardous waste sources, 16
  NPL sites with natural attenuation as part of final remedy, 212

Z

Zinc, 165
  data needs for natural attenuation, 154
  hazardous waste sources, 16, 20
  likely natural attenuation pathways, 153
  solubility, 49
  uranium mill tailings, 106